THE EROTIC MIND

THE EROTIC MIND

Unlocking the Inner Sources of Sexual Passion and Fulfillment

JACK MORIN, Ph.D.

HARPER

NEW YORK • LONDON • TORONTO • SYDNEY

HARPER

A hardcover edition of this book was published in 1995 by HarperCollins Publishers.

THE EROTIC MIND. Copyright © 1995 by Jack Morin, Ph.D. All rights reserved. Printed in the United States of America. No part of this book may be used or reproduced in any manner whatsoever without written permission except in the case of brief quotations embodied in critical articles and reviews. For information, address HarperCollins Publishers, 195 Broadway, New York, NY 10007.

HarperCollins books may be purchased for educational, business, or sales promotional use. For information, please e-mail the Special Markets Department at SPsales@harpercollins.com.

First HarperPerennial edition published 1996.

Designed by Nancy Singer

The Library of Congress has catalogued the hardcover edition as follows:

Morin, Jack.

The erotic mind : unlocking the inner sources of sexual passion and fulfillment / Jack Morin — 1st ed.

p. cm.

Includes bibliographical references and index.

ISBN 0-06-016975-3

1. Sexual excitement. 2. Sex (Psychology). I. Title.

HQ21.M7945 1995

155.3'1—dc20 95-4944

ISBN 0-06-098428-7 (pbk.)

23 24 25 26 27 LBC 36 35 34 33 32

HB 11.21.2023 0438

To the men and women who revealed their erotic secrets so that others might be inspired to explore their own

CONTENTS

ACKNOWLEDGMENTS

Once in a while I hear of a book moving from conception to fruition with enviable rapidity and ease. *The Erotic Mind* has never been such a project. More than a dozen years have elapsed between the first realization that I had to write this book and its eventual completion. I only hope that the many phases of my work have ultimately added to its depth and usefulness. This much I know: without the encouragement, steadfast support, penetrating insights, and constructive criticisms of a remarkable group of people, I would never have persisted. So it's especially gratifying to be able to thank them now.

I'll forever be grateful to my clients who courageously examined their eroticism with me in psychotherapy. I also feel profound appreciation for the hundreds of people I don't even know who responded to my anonymous survey about peak erotic experiences. The willingness of clients and respondents alike to reveal what most people keep hidden made it possible for me to develop a new way of understanding erotic life.

As my ideas crystallized I received sustenance from colleagues and friends who seemed to grasp immediately the implications of my viewpoint and pushed me to take it further—especially Toni Ayres, Joani Blank, and Michael Graves, all of whom also gave invaluable assistance in the development of the Sexual Excitement Survey. In addition, Michael focused his impressive analytical skills on at least two versions of the entire book and contributed greatly to its final structure.

I also benefited from frequent talks with Marty Klein about the intricacies of both writing and sexuality. Janice Epp, LouAnne Cole, and Gary Zinik probably don't realize how much I've valued their ongoing resonance with my work. Gary also helped me work out some tricky aspects of thematic analysis. And Lonnie

Barbach read much of the manuscript and offered many helpful suggestions.

I found a consistently challenging forum for presenting my ideas at meetings of the Society for the Scientific Study of Sex, an international association of sexologists. Questions, critiques, and long conversations with dozens of colleagues influenced my work more than they can possibly know. Their interest buoyed me, especially during several years of struggle with an unrelenting case of writer's block. Had I been unable to articulate my thoughts and findings in these lively discussions, I might have given up.

It was Hal Bennett, a talented writer, editor, and book consultant, who helped me resolve my blockage with his gentle wisdom. He also showed me how to shape complex material into a cohesive, accessible whole. When I ran into snags, he always knew how to nudge me back on track. Then Hal introduced me to Fred Hill, who became my agent. He was another find, a fact I realized when I overheard him describing the book far more succinctly than I could have done myself.

Fred thought he knew exactly who would want to publish the book: HarperCollins's editor-in-chief, Susan Moldow. Indeed she did and proceeded to perform editorial magic with uncommon understanding and respect for an author's sensitivities. She also selected Nancy Nicholas, a gifted line editor from whom I learned so much about unnecessary words. As we were nearing the final round of revisions, Susan left HarperCollins and entrusted the book to Gladys Carr and Cynthia Barrett, both of whom made an eleventh-hour transition that could easily have been a nightmare into an opportunity. Their suggested refinements were right on the mark.

Throughout it all I was blessed with cherished friends and family who, miraculously, maintained enthusiasm for my work in spite of the fact that I was often unavailable or preoccupied. My dearest friend, Scott Madover, both gave and endured the most. I doubt I can ever repay him adequately.

Jack Morin, Ph.D.
San Francisco

Introduction

SEX AND SELF-DISCOVERY

Appreciating the mysteries of eros requires a new point of view.

Ever since the sensation caused by Alfred Kinsey's monumental studies of sexual behavior, we seem to have developed an insatiable appetite for this kind of information. It's not simply that we're voyeuristic or enjoy being titillated and entertained—although most of us do. Many of us also crave reassurance that we're sexually "normal," hope to resolve sexual problems, or look for ways to enhance our enjoyment. Others are wondering how to keep the spark of passion alive in long-term relationships or marriages.

Never before have so many had so many opportunities to learn so much about sex. And yet the most important aspects of the sexual experience are the ones we know least about and are most reluctant to discuss. For example, we know far more about which sexual acts people engage in, how often, and with whom, than we know about what makes some of these acts compelling. We are awash in facts and figures, yet relatively little has been written about what sex means for those involved—such as how it enriches their lives or helps them feel better about themselves.

The greatest gap of all exists between our extensive knowledge about the physiology of sex and our relatively rudimentary awareness of the psychology of arousal—how our minds create, intensify, or restrict sexual enthusiasm. Considering how important these issues are for a satisfying sex life, our sketchy understanding here is nothing short of tragic. Sex therapist Bernie Zilbergeld puts it bluntly:

> More than anything else, arousal is what drives good sex. It *is* the spark. It is also the cornerstone of a sexuality based on pleasure rather than on performance. If you want more exciting and more satisfying sex, go for greater arousal.[1]

Almost everyone is aware that becoming aroused feels good. But how can we create greater arousal when we know so little about it? Why are certain people, images, and situations so much more stimulating than others? Why do individual preferences and patterns vary so dramatically? Why are most of us attached to specific turn-ons? And what do our turn-ons reveal about who we are and what we're searching for?

These and other unanswered questions about arousal lie at the heart of this book. To further our understanding we must broaden our interest beyond mere sex and explore the world of eroticism. The erotic landscape is vastly larger, richer, and more intricate than the physiology of sex or any repertoire of sexual techniques. The more mechanical and explicit aspects of sex are relatively easy to observe and translate into numbers and graphs, whereas the most rewarding and powerful secrets of eroticism are elusive, highly individualistic, and difficult to quantify. To make sense of it we must cultivate a whole new way of perceiving.

WHAT IS EROTICISM?

The modern concept of eroticism is rooted in Greek mythology. Eros was the young and playful god of love, son of Aphrodite, the goddess of love and beauty. At the most basic level Eros is the source of attraction and the craving for sexual love. Some Greek philosophers also saw Eros as the force behind all creation, his absence leading to decay and destruction. Sigmund

Freud was drawn to this interpretation, associating eros—which he conceived as energy rather than a god—with his concept of libido, a combination of sexual drive and life force.*

Eroticism can best be understood as the multifaceted process through which our innate capacity for arousal is shaped, focused, suppressed, and expressed. We're born sensuous and sexual, but we *become* erotic as we receive both overt and subtle messages about ourselves from our primary caretakers and gradually integrate these messages with our experiences of touch, as well as the highly personal mental images and emotions that go with them. As we grow, the demands and ideals of our culture, along with the interpersonal dynamics of our families and communities, influence our responses profoundly.

Eroticism is the process through which sex becomes meaningful. By the time we reach adulthood we've all discovered that, by itself, sex can be little more than a collection of urges and acts. But the erotic is intricately connected with our hopes, expectations, struggles, and anxieties—everything that makes us human. Whereas sex can be simple, by its very nature eroticism is complex, and from this richness true passions are born. It is also through the magic of eros that sex and our search for emotional closeness become intertwined.

Sexologist David Schnarch defines eroticism as "the pursuit and delight in sexual pleasure."[2] This beautifully idealistic definition certainly describes those notable moments when we are able to express and celebrate our erotic selves without restraint. But it leaves out the fact that eros is energized by the *entire* human drama, including the unruly impulses and painful lessons that no one—except those who retreat from life—can possibly avoid. No wonder the erotic mind conjures up images of debauchery as well as delight. Because it is connected with all aspects of existence, I define eroticism as *the interplay of sexual arousal with the challenges of living and loving.*

---

*Throughout this book I use "eros" in this broader sense rather than as the name of a god. Therefore, I do not capitalize it.

A PARADOXICAL POINT OF VIEW

Your ability to probe the mysteries of the erotic mind insightfully—and to benefit from your discoveries—will be enhanced enormously if you consciously cultivate a stance toward the erotic that allows you to appreciate fully its rich complexity. The two predominant schools of thought in contemporary psychology—which I call the pathological and the neat-and-clean—have unwittingly limited our understanding of eroticism at least as much as they have advanced it. These two viewpoints extend way beyond psychology, permeating modern attitudes in countless ways.

EMPHASIZING PATHOLOGY

The pathology perspective is by far the older and more deeply entrenched. Freud and his psychoanalytic followers have always been the most articulate proponents of this perspective. From its inception psychoanalysis has been very interested in eroticism, largely as a result of Freud's two great contributions to the study of sex: (1) making us aware of the profound influence the unconscious has in shaping personality and therefore eroticism and (2) forcing us to recognize that children are sexual beings—although not, of course, in the same way as adults. Freud rightly insisted that sexuality and personality unfold in tandem, with much of that process occurring outside conscious awareness.

Whereas psychoanalysts have done us all a service by recognizing and investigating the untamed powers of eros, most make their observations through a dark lens. To a far greater degree than today's theorists of pathology are aware, the pathology model of sex is rooted in the concept of sex-as-sin. Almost like a fire-and-brimstone preacher, Freud himself believed that, if left unchecked, erotic urges would wreak havoc.

The psychological word for sin is "perversion," and in some circles it's shockingly easy to qualify for that label. Those who are attracted to the "wrong" people, who eroticize inanimate objects, who have the "wrong" kinds of orgasms, or who express "infantile" preferences—such as strong interests in masturbation, oral, or anal stimulation—have, at one time or another, all been stigmatized by the pathology approach to sex. Pathology-oriented psychologies typically propose such a narrow path for "normal" erotic development that hardly anyone ever attains it. Notions of erotic health are often little more than abstractions derived almost entirely from theory and from therapeutic work with those who are sexually troubled. While a great deal can be understood through theory and therapy, unless we also study eroticism in its most positive forms, many of our conclusions are bound to be distorted.

SANITIZING EROS

The neat-and-clean perspective is a more recent development based primarily on the precepts of behaviorism and humanism. Behaviorists see human sexuality as governed by predictable rules of learning. They are convinced that the unruly tendencies of eros can be controlled by a little tinkering with rewards and punishments.

Humanistic psychologists, fed up with the traditional focus on pathology, emphasize human potential and our capacity for growth. This approach allows people much more freedom to be human, and thus diverse. But the humanists have shockingly little to say about sex, and even less about eroticism. They prefer to talk about love and intimacy, worthy topics to be sure, but made strangely one-sided by ignoring our lustier impulses. I'm humanistically oriented myself and have always been perplexed about how eroticism became the neglected stepchild of humanistic psychology. Could it be that it's just too messy?

Masters and Johnson launched the field of modern sex therapy in 1970.[3] They started with the traditional medical model with an emphasis on physiological function and dysfunction. Then from humanism they incorporated an acceptance of sexual variations and an optimistic belief that people gravitate toward health. Using principles of behaviorism they developed practical, step-by-step methods for facilitating change. This mixture proved to be far more effective than traditional approaches in helping people resolve problems with their sexual functioning.

Like all new fields, modern sex therapy has limitations, one of which is the fact that it is firmly rooted in the neat-and-clean perspective. Masters and Johnson considered sex to be fundamentally simple—a "natural" function. According to this view, all that is required for a happy sex life is the removal of disruptive impediments, most notably anxiety and guilt. If it ever occurred to them that the same impediments they sought to eliminate might also be turn-ons, they neglected to mention it.

Many of today's sex therapists are still so preoccupied with getting rid of inhibitions that they rarely stop to wonder how sexual passion is actually created and intensified. They assume, for example, that uncomplicated, comfortable sex is the best kind—which, of course, isn't necessarily so. Neat-and-clean practitioners also believe that the ability to become aroused and to have orgasms should be sufficient for satisfaction. They don't dwell on the fact that eroticism is intertwined with the untidy struggles of being human and is therefore inherently complex and unpredictable.

EMBRACING PARADOX

This book is based on a completely different point of view. Whereas the pathology perspective fails to appreciate the inevitable variability of eroticism, and the neat-and-clean perspective tries to downplay its irrational power, the paradoxical perspective recognizes the joys of eros without denying its intricacies and risks. This new paradigm acknowledges and embraces the contradictory, dual-edged nature of erotic life. It recognizes that anything that inhibits arousal—including anxiety or guilt—can, under different circumstances, amplify it. Consequently, we must view with considerable skepticism any absolute statements about what makes sex either exciting or problematic.

Many find it discomforting to tolerate the ambiguity of the erotic experience, to accept its mixed motivations, or to observe how the erotic mind has a habit of transforming one idea or emotion into another. And yet if we fail to come to terms with the fundamentally paradoxical nature of eroticism, we set the stage for its negative aspects to appear more frighteningly destructive. At the same time, the positive aspects of erotic life become increasingly elusive and difficult to celebrate, almost as if they are canceled out by a recognition of the danger or uncertainty inherent in them.

The paradoxical perspective is the best alternative to the exaggerated shifts in sexual attitudes we've seen during the last few decades. According to the ethos of the 1960s and 1970s, sexual experimentation and freedom were valued. For millions it was a time to throw off old restraints, to push boundaries, and sometimes to overindulge recklessly.

Within a single decade the pendulum lurched back in the opposite direction. In response to fear of AIDS and other sexually transmitted diseases, as well as the atmosphere of conservative politics, social attitudes about sex flipped from celebration to dread. Reflecting and abetting this radical shift, the popular media turned its attention to the most disturbing manifestations of the erotic impulse, including sexual abuse and harassment, teen pregnancies, disease, and even satanic cults. In record time the popular perception of sex went from "good" to "bad." No wonder so many of us are confused and a bit dazed.

We live in an era of both promise and great danger. The danger is that negativity will drive eroticism into the shadows where it is most likely to assume the very shapes we fear. But those who find the courage to survey the entire panoply of the erotic experience—joyful as well as dangerous, life-giving as well as troublesome—stand on the brink of a new consciousness of eros. As our perspective enlarges we can see that, in the final analysis, eroticism can never be either pathological or neat-and-clean—for it is as vast and multifaceted as human nature itself. The paradoxical perspective is the only point of view large enough to encompass this truth.

WHAT THIS BOOK CAN DO FOR YOU

The Erotic Mind is an invitation to a deeper understanding and appreciation of the role of eroticism in your life. The book is divided into three parts. Part I, "Realms of Passion," focuses primarily on peak erotic experiences as rich sources of information about the inner workings of the erotic mind. Particularly memorable real-life encounters as well as compelling fantasies offer us glimpses of eroticism thriving—as opposed to malfunctioning or causing trouble. In addition, during moments of peak arousal, the dynamics of eroticism are accentuated and thus easier to observe.

I first realized the value of peak sexual experiences in my work as a therapist. Years later I created the Sexual Excitement Survey so that respondents could write anonymously, and in considerable detail, about their most unforgettable turn-ons. Their sexy tales will serve as fascinating illustrations of the erotic mind in action.

If you read this book solely as an interested observer, I'm sure you'll be informed and entertained. But I hope you'll use it to embark on a voyage of erotic self-exploration. I will regularly invite you to examine your own peak erotic experiences and show you how to search gently for clues to your eroticism. You'll open new pathways to sexual satisfaction by reading about other people's erotic truths—the ones that rarely are given a voice—while simultaneously attending to your own.

You'll identify the factors that make peak experiences stand out and discover that eroticism is dynamic and paradoxical because it springs from the interplay between your attractions and the obstacles that stand in your way. It will become clear that predictable challenges of early life, faced by us all, become the cornerstones of eroticism. You already know how positive emotions can energize arousal, but you'll see how unexpected aphrodisiacs such as anxiety, guilt, and anger can have similar effects. As your awareness expands you'll marvel at your erotic mind's amazing ability to transform life's inevitable difficulties and emotional wounds into sources of excitation.

In Part II, "Troublesome Turn-ons," we'll use the paradoxical perspective to gain insight into extremely common but rarely discussed erotic problems. We'll look at how the same emotions that excite us can turn against us, blocking our enjoyment. We'll also explore how ingrained erotic patterns can unconsciously draw us into relationships as frustrating as they are intense. Then we'll discover how erotic conflicts during childhood and adolescence can split our lustful from our tender feelings. And we'll observe how low self-esteem sometimes links with high arousal to produce the most overwhelming and destructive of all turn-ons. As you study these difficulties they will become understandable rather than frightening. Part II concludes with a practical look at the erotic mind as a potent force for healing and growth, including a seven-step program that anyone can follow to promote positive erotic changes.

In Part III, "Positively Erotic," we'll explore how your growing self-awareness can guide you toward more joy and satisfaction in your life. The paradoxical perspective will provide you with new insights into the age-old challenge of deepening intimacy while sustaining passion in long-term relationships. While acknowledging the inherent conflict between closeness and sexual excitement, you'll learn how erotic couples actively cultivate key skills to keep the spark alive between them as their relationships grow. Next we'll reevaluate the true nature of erotic health. Crucial signposts will help you evaluate your sexual well-being and prepare you for lifelong erotic development. The book concludes with an inspiring look at the rewards of the erotic adventure.

A WORD ABOUT AIDS AND SAFE SEX

It's difficult to imagine anyone who's unaware of the international scourge of AIDS. All of us, regardless of gender or sexual orientation, must evaluate our sexual behavior to minimize the risk of exposure to HIV, the virus that causes AIDS. There's disagreement about the odds (seemingly quite low) of contracting HIV through activities such as tongue kissing or oral sex without ejaculation. But the centerpiece of any safe sex policy is the decision never to engage in vaginal or anal intercourse without using a latex condom—or to abstain altogether.

The only sexually active people who needn't be concerned are couples who have been monogamous for at least six months and have both recently tested negative for the antibodies to HIV, indicating that they've never been exposed. As long as these couples can trust each other to avoid unsafe sex outside the relationship, HIV isn't an issue.

Keep in mind that AIDS prevention is not covered in this book. You'll notice stories of peak sex involving intercourse in nonmonogamous situations that don't mention condoms. Of course, some of these encounters occurred before the AIDS epidemic. Other people may have used condoms but didn't say so because it was merely a practical consideration (like birth control) that, although important, didn't play a role in making their encounters memorable. Some people undoubtedly took unwise risks. Fantasies, of course, are totally exempt from all practical considerations. If you have questions about AIDS and how to prevent it, ask your local health department, an organization concerned with AIDS, or your physician.

ATTITUDES FOR SELF-DISCOVERY

Having had the privilege of guiding hundreds of men and women on voyages of erotic self-discovery, I've consistently observed that venturing into the erotic realm is dramatically more rewarding for those who, right from the start, make a point of learning how to (1) suspend judgments, (2) trust themselves, and (3) use a gentle approach.

SUSPENDING JUDGMENTS

My most fervent belief about erotic self-discovery is that deep erotic truths do not reveal themselves to judgmental eyes and nonaccepting hearts. The secrets of the erotic mind become visible only to those who consciously cultivate the ability to set aside critical evaluations long enough to perceive what's actually there. Until you learn this all-important skill, your perceptions will be skewed and their usefulness dramatically reduced.

The most emotionally charged judgments arise from moral convictions. Not surprisingly, those who believe in immutable moral doctrines usually have the most difficulty setting aside their automatic inclinations to reject or criticize actions and thoughts that violate their concepts of right and wrong. But not all judgments are rooted in morality. Many people experience a similar visceral discomfort when their erotic experiences don't match their expectations, or when they realize that what truly excites them conflicts with their ideals.

The only way to learn consciously to suspend judgments is through practice in a nonthreatening environment. Just like children, when we fear being judged or ridiculed—by ourselves or others—we become immobilized or take refuge in rigid beliefs and habits. But any observer who suspends judgment, even for a moment, immediately becomes more open-minded and relaxed.

How can you develop this skill? The first step is to begin acknowledging what triggers your critical responses. Then you must decide whether to repeat those responses or to try setting them aside. All that matters now is that you become aware of when you're making judgments—especially if they occur automatically and are therefore invisible.

Some people have a mistaken notion that suspending judgments means adopting an amoral stance toward sexuality or negating the importance of one's values. Nothing could be further from the truth. In fact, only through a courageous examination of the difficult truths of erotic life does it become possible to establish a meaningful ethical system to guide our actions.

TRUSTING YOURSELF

Few of us have been taught that we can be trusted in the erotic realm—quite the opposite. As youngsters most of us were encouraged by a variety of subliminal and explicit messages to be wary of our eroticism as it was developing. Understandably, adults who have learned these lessons well often feel squeamish about examining the content and meaning of their turn-ons. It's not surprising or unusual to harbor a semiconscious concern that you might uncover things about yourself that you would be better off not knowing.

You can imagine how such an attitude restricts your vision. But if your childhood training has made you uneasy about eroticism, denying that attitude will only make matters worse. The best thing you can do is to acknowledge your feelings, no matter how illogical it may seem, and avoid putting yourself down for them. Discomfort with one's sexuality takes years to build up and can't be changed overnight.

Your journey of erotic self-discovery will be infinitely easier and more rewarding if you can find and nurture even a small spark of faith in yourself. As you come to realize that eros is fueled by the energy of life itself—and thus contains a deep-rooted urge toward growth and self-affirmation—the legacy of mistrust can be gradually overcome. Keep in mind that those who establish a comfortable acceptance of their erotic urges are the least likely to inflict harm upon themselves or others.

USING A GENTLE APPROACH

Although many have tried to command the inner secrets of their erotic life to reveal themselves, no one has succeeded. Your erotic mind, fearing condemnation or rejection, has become adept at concealing itself. Because few of us are free to express our unfolding eroticism openly, hiding the truth—even from yourself—begins as an act of primitive self-preservation.

To uncover what has long been hidden, be patient and gentle; allow the erotic mind to reveal itself at its own pace as it tests the waters. Practice offering yourself invitations to see more, to comprehend more, to accept more, to enjoy more. Each invitation carries with it the freedom to decline or to wait. The goal of erotic self-understanding is furthered by a willingness to ease up in the face of your own reluctance.

These three attitudes taken together—suspending judgments, trusting yourself, and using a gentle approach—help bring you face to face with eros in action without being afraid. Armed with an appreciation of paradox, a willingness to venture boldly into uncharted territory, and a sense of awe and wonder, you are ready to uncap the wellsprings of passion. And what of fulfillment? You will discover that passion and fulfillment are intricately linked yet distinct experiences. Just as surely as passion strives for fulfillment, fulfillment longs for passion. Between the two, eros flourishes.

Part I

REALMS OF PASSION

1

PEAK EROTIC EXPERIENCES

Unforgettable turn-ons are windows into your erotic mind.

One of the most effective and enjoyable ways to unlock the mysteries of the eros is to reminisce about your most compelling turn-ons. During these moments of high arousal the crucial elements—your partner, the setting, perhaps a tantalizing twist of luck—all mesh like instruments of an orchestra, producing a crescendo of passion. Look closely at a peak turn-on and you'll undoubtedly sense that something close to the core of your being has been touched. And because everything is accentuated during such moments, they reveal an enormous amount about how your eroticism works.

As a young psychology student in the 1960s I was influenced by Abraham Maslow, who called for a "psychology of health" to counterbalance the overemphasis on problems that he believed was distorting our view of human beings. He broke new ground by studying people he called "self-actualizers"—those who are comfortable with themselves, relatively free of neurotic conflicts from the past, and available to tackle the challenges of living with creativity and zest. Self-actualizers are still largely ignored by psychologists, even though they have much to teach us about emotional well-being.

Maslow was equally intrigued by a wide variety of *peak expe-*

riences—such as being enraptured by a beautiful piece of music or a painting, a special communion with nature, or the joy of bodily expression in dance or athletics, to name just a few.[1] During these moments of ecstasy we are fully present in the moment, unself-consciously expressing our truest selves with ease and grace, grateful to be alive. Even though peak experiences aren't "productive" in the usual sense, participants invariably describe them as profoundly positive and sometimes even life-changing.

According to Maslow, self-actualizers have peak experiences more frequently than the rest of us, but nearly everyone has them occasionally. Among his most provocative observations was that during and following peak experiences we temporarily take on many of the characteristics of self-actualizers. In other words, peaks offer us glimpses of our most authentic, healthiest selves and thus can serve us as guides to growth. Maslow saw peak experiences as crucial sources of "clean and uncontaminated data" about who we are and might become.[2]

When I began my formal studies of eroticism as a practicing psychotherapist I approached the challenge with Maslow's insights in mind. I was convinced that if I devoted as much attention to peak sexual experiences as I did to problems, I could eventually discern truths about eroticism that would otherwise elude me. My first discovery was rather discouraging: even in the nonjudgmental atmosphere of therapy people rarely bring up their peak turn-ons spontaneously. And when I started asking I quickly learned that most clients required a high comfort level and a significant amount of courage before they were willing to disclose details about this extremely intimate material.

I began encouraging clients who were grappling with sexual problems to explore their peak turn-ons, hoping the potential benefits of doing so would be obvious to them. In most cases I was wrong. The majority had trouble grasping the value of discussing their peak experiences; they just wanted to fix their problems. A prevalent comment was, "Sure I've had good sex in the past but what can that do for me now?" Out of necessity I became adept at gently challenging clients to set aside their preoccupation with problems for a while so they might learn more about their eroticism.

I quickly saw that those who accepted my challenge typically made more rapid and long-lasting progress than those who insisted on focusing exclusively on their troubles. Some improvements came about when they used their peak turn-ons to help clarify their conditions for satisfying sex—an extremely important ingredient for successful sex therapy.

Fred: Centerfold syndrome?

Fred consulted me because his sexual desire for Janette, his wife of six years, had been declining for more than a year. Although he assumed she must have noticed the reduction in both the frequency of sex and his enthusiasm, Fred had no idea how to discuss his predicament with Janette without hurting her. Besides, he felt ashamed of himself and was convinced she couldn't possibly understand what he was going through.

"I think I have the centerfold syndrome," he announced about halfway through our first meeting. He explained that as a young adolescent he had masturbated to photos in his dad's *Playboy* magazines. More than twenty years later the majority of his fantasies were still populated by young women with picture-perfect bodies. "My wife still looks great," he added, "but she's no centerfold. I know it sounds horrible to say but I can't help noticing her body changing. I love her too much to tell her the truth."

Clients, especially introspective ones like Fred, often enter therapy with theories about the origins of their problems. Fred was the first, however, to have invented and named a new diagnosis! Yet many other men—and more than a few women too—had hinted that constant images of sexual perfection in the mass media sometimes reduced the allure of their actual partners. Obviously, Fred had named a very real problem.

However, as Fred talked about his sex history and his relationship with Janette, I sensed that his declining desire had less to do with flawless centerfolds than he believed. To help him find out for himself I suggested that he think about his peak turn-ons. After overcoming his initial hesitation he told me about a series of memorable encounters with a young waitress and aspiring model with whom he had an affair when he was in the military. His obvious pleasure in telling these stories soon turned to discouragement because they seemed to confirm his theory. I suggested that before jumping to conclusions he consider other factors that also might have turned him on besides her gorgeous body.

As he revealed the details of these remarkably passionate encounters it became clear that in each instance his excitement reached its zenith when the waitress unambiguously demonstrated her attraction for him by "losing control" and "going after what she wanted." Certainly her appearance stimulated his desire, but it was her escalating enthusiasm that transformed simple desire into fiery passion. Fred discovered that when he fantasized about a centerfoldlike woman the visual enjoyment of her perfect body was merely a starting point. As his arousal built he imagined her casting caution to the wind and "going crazy with lust."

Then I asked about peak encounters with his wife, of which their had been many, even some during the previous year. Because he had already recognized how much he valued a highly responsive and eager partner he quickly saw that in every peak encounter with Janette, she too had been unusually expressive and uninhibited. Fred also stumbled on another important fact: whenever Janette exhibited the impassioned fervor that excited him, he automatically *perceived* her differently; he completely ignored her imperfections and focused instead on her most appealing attributes.

As it turned out, Fred's declining desire was influenced—but not caused—by his fascination with centerfolds. Much more important than Janette's changing body was the fact that she had become increasingly passive in recent years, to the point where Fred was no longer sure if she particularly enjoyed sex with him anymore. Luckily, the two of them had always talked openly about most things, although it had been years since they had discussed sex. Spurred on by a deeper understanding of his eroticism and a reduction of his guilt, Fred initiated an extremely productive dialogue with Janette.

Once she felt assured that his intention was not to criticize her but to improve their sex life, she acknowledged her increasing passivity in bed. She explained how Fred's tendency to be sexually aggressive had led her to conclude that he *preferred* her to be submissive. She was just trying to give him what she assumed he wanted. She also divulged that her passivity was making sex less interesting for her too. As she grasped how much Fred missed feeling her unbridled desire, she gradually felt freer to let her excitement show. And in response, Fred's passions once again began to stir.

Fred might have found a solution to his problem without examining his peak turn-ons. And, of course, this couple's ability to communicate so productively and become sexually experimental—skills that many sex therapy clients lack—were essential to their success. But by freeing him from the discouraging and mistaken belief that only physical perfection could excite him, Fred's memories of peak sex greatly accelerated his progress.

As I saw the benefits of investigating peak turn-ons in dozens of sex therapy cases, including ones much more difficult than Fred's, I wondered if similar memories might also help with problems not directly linked to sex. Soon I had an intriguing opportunity to find out.

Sabrina: Rediscovering vitality

After she had struggled with "blue moods" for several years, Sabrina's depression was getting worse. "Nothing matters that much anymore," she lamented. "I'm lonely most of the time. I've forgotten how to have fun. I feel ugly. And I'm not even sure my husband loves me." With that she burst into tears. It was easy to see she was suffering from a "love depression" based on the belief—shared by all the women in her family—that men are incapable of loving women, and therefore she was doomed to a life of dissatisfaction. Speaking of her husband, Ted, she complained, "If he's not working, he's puttering in the damn garden. He hardly ever notices me, let alone holds or kisses me."

She was surprised when I asked if she could remember any especially fulfilling sexual encounters with Ted. Even though she couldn't see how my question was relevant to her depression, she halfheartedly agreed to consider it during the week. At the beginning of our next session she pulled a crumpled piece of paper from her purse. After an awkward silence, followed by a deep sigh and then a slight smile (the first I had seen), she read a story complete with spontaneous commentary:

> Like most people who have been married nineteen years, Ted and I have a lot of routines, including making love on Saturday mornings before starting our chores. This tends to be rather mechanical and obligatory. One Saturday—I think it was about three or four years ago—I woke up late and Ted was already out in the garden. I opened the drapes to a glorious morning. There was Ted in his overalls, digging in the dirt, whistling. Instead of feeling hurt that he was ignoring me, I thought how cute he looked and how happy I felt. Even then that was pretty rare because I was often pissed off with Ted for not showing me enough affection.
>
> It's so silly what happened over the next several hours. [She made me promise not to laugh.] We were both so *different*. I joined him in the yard and we instantly began flirting. He made sexy comments under his breath. I remember one: "I'd like to rub my face in that bush over there," motioning toward me. I tried a few innuendoes myself, nutty stuff like, "Is that a trowel in your pocket or are you just glad to see me?"
>
> He made me feel so free and sexy that I unbuttoned my shirt. When he saw my breasts he instantly dropped what he was doing and began licking my nipples and smearing dirt all over me. I can't explain why this didn't bother me because I'm such a cleanliness fanatic. Before long we were both stark naked, fighting over who would pull each weed, sometimes rolling on top of each other, laughing our heads off, and being totally outrageous. The sun was hot so I grabbed the hose and sprayed him down. Soon we were both drenched and making love on the lawn.
>
> The most exciting part was when Ted gazed into my eyes and said with so much feeling I was absolutely overwhelmed, "I love you more than anyone in the whole world."

"I have no idea what got into him," Sabrina added, "but I sure wish Ted would be like that more often!"

"And what about you?" I asked. "What if *you* were like that more often?" Sabrina didn't care for that question one bit, which she demonstrated by wiping the radiant smile off her face and stuffing her joyous tale back into her purse. She remained mostly silent for the rest of the session.

Depression makes one feel dull, lifeless, and helpless. In stark contrast, peak eroticism always fosters energy and vitality. Sabrina's lively story couldn't have been more out of step with her chronic blue moods, and I sensed she was struggling with that contradiction. I was pleasantly surprised when she brought up her story again the following week. "You think *I* had something to do with Ted and I acting so differently that day, don't you?" The conviction in her voice told me she knew she had.

Over many weeks Sabrina catalogued how she had somehow set aside her usual ways of thinking, feeling, and acting during that peak experience. *She* interpreted Ted's working in the garden as endearing rather than a sign of rejection. *She* allowed herself to be moved by the beautiful morning. *She* actively participated in creating a playful atmosphere. And most important of all, *she* seized the opportunity to become vibrantly erotic. Gradually, Sabrina embraced her peak experience as evidence of what could happen if she stopped clinging to her lonely fate and recognized her abilities to make things different. She read the story to Ted and taped it to her bedroom mirror as a reminder.

Anyone who's ever been seriously depressed knows how difficult it is to move beyond feelings of helplessness. Sabrina's situation was complicated by the fact that she had selected a husband who matched her expectations; he often *was* emotionally distant and uncomfortable with closeness. Yet as Sabrina slowly reconnected with her vitality she became more approachable—and a lot more fun to be with. Sometimes Ted responded positively, and she would practice taking in his affection without critiquing it. When he was preoccupied and unavailable she learned to separate her mood from Ted's personality quirks. Sabrina saw welcome improvements in their marriage and sex life, although neither was perfect. She did, however, cultivate a more active stance in her world, which made her far less despairing.

Fred used his peak turn-ons to discover new *information* about his eroticism, whereas Sabrina used hers as *inspiration*. I have found these to be the two most common therapeutic benefits of exploring peak eroticism. But anyone who takes the time to examine the nuances of peak turn-ons will gain valuable insights into how the erotic mind creatively expresses our innermost needs and potentials. This knowledge, however, doesn't necessarily come easily.

The more deeply we explore, the more we see that peak turn-ons are not only very similar to other peak experiences, they're also different. Whereas most peak experiences contain little to be embarrassed about, in peak sex the erotic impulse frequently strays far from our ideals. In the realm of eros all the contradictions and paradoxes of the human drama are played out. One way or another, erotic peaks always reveal secrets about our idiosyncracies, conflicts, and unresolved emotional wounds. More often than not, people fear that if their innermost experiences of arousal are revealed they will be pronounced abnormal. No wonder we discuss such matters, if we do so at all, only when we feel safe and are absolutely certain that we won't be judged or ridiculed—by others or ourselves.

THE SEXUAL EXCITEMENT SURVEY

In the mid-1980s I realized that even though I was learning an incredible amount about the erotic mind in my therapeutic work, my studies were being hampered. For one thing, relatively few clients were as open about their eroticism as Fred and Sabrina. In addition, most clients who did explore their peak experiences were impatient to get back to the problems that had brought them to therapy in the first place. I became keenly aware of both the advantages and the limitations of therapy as a means of investigating eroticism. I was eager to expand my work by studying peak erotic experiences in a totally different way.

To that end I created the Sexual Excitement Survey (SES). The survey asks anonymous respondents to write in detail about especially arousing and memorable encounters and fantasies, as well as their ideas about what made these events so thrilling. My challenge would be to analyze the content of their stories and comments and look for recurring themes and patterns.

It's difficult to find people willing to spend at least ninety minutes disclosing the very things they naturally keep to themselves. So I distributed many of the surveys in undergraduate- and graduate-level human sexuality classes, where self-exploration is a part of the learning. Interested students mailed completed surveys directly to me—not to their instructors. Also, a number of professional and social organizations took an interest in this project and invited their members to participate. Whenever I spoke in seminars and workshops, I always mentioned the SES and had a stack of them available.

MEET THE GROUP

Over a period of two years 351 respondents—whom I affectionately call The Group—accepted my invitation to reveal the most intimate details of their eroticism. Collectively, they described 687 memorable encounters and 339 favorite fantasies, for a total of more than 1,000 peak erotic events. Compared to the tens of thousands of respondents who regularly respond to questionnaires in popular magazines, these are obviously small numbers. But the SES is not a typical survey because of the depth of self-disclosure it requests. Instead of learning a little bit about thousands of people, I preferred to learn a great deal about a few hundred.[3] Through my analysis I intended to become intimately acquainted with every one of these strangers.

Fortunately, The Group is as diverse as I had hoped.* It consists of men and women ranging in age from eighteen to sixty-nine, representing many races, all sexual orientations, and a wide variety of types and frequencies of sexual behavior.[4] They differ from the general population primarily in their willingness to

---

*Although The Group obviously responded to the SES in the past, their revelations form a body of data that exists perpetually in the present. Therefore, I always refer to The Group in the present tense, even when they use the past tense to describe their experiences.

write about such matters, as well as their inclination toward introspection, well-developed verbal skills, and relatively high levels of formal education.

RECALLING PEAK ENCOUNTERS

Most sex surveys are written to benefit the researchers. Although I created the SES to help answer questions that matter to me, from the beginning I wanted the survey to give something back to the respondents. Members of The Group frequently mention how they personally benefited from filling out the SES. I'm also delighted that many therapists now suggest that their clients use the SES as a consciousness-raising tool.

To get the most out of this book, I invite you to contemplate the same questions that I asked The Group. You'll find the SES, along with a simple set of instructions, in the appendix. Consider responding to the entire SES before reading any further to be sure your answers are completely spontaneous. Once you complete the SES you might wish to send me your answers (*without* your name, of course) so that I can expand my research to a larger population. That decision, clearly, is completely up to you and can be made at any time.

If the SES seems a bit daunting just now, an alternative is to take it one step at a time. Start with two key questions about your most memorable real-life encounters:

1. Think back over all your sexual encounters with other people. Allow your mind to focus on *two* specific encounters that were among the most arousing of your entire life. Describe each of them in as much detail as you wish.
2. What are your ideas about what made each of these encounters so exciting?

As you scan your memory you'll probably recall a variety of experiences. Keep in mind that peak encounters are not necessarily dramatic or sensational. Sometimes the best ones are remarkably simple. Nor is it necessary for peak encounters to include intercourse—or any particular sexual act. In a number of The

Group's marvelous stories the participants are fully clothed and some of the stories are exclusively visual, without any touching at all! Although the SES asks for only two peak encounters, there's no reason to limit yourself. Feel free to recollect as many as you wish.

If you haven't yet had any erotic experiences that you consider peaks, concentrate on those that were the most appealing so far. If you haven't had *any* sexual encounters yet, focus on feelings of curiosity, desire, or attraction that stand out. It might also be helpful to pay special attention to sexual fantasies, which we'll be discussing shortly.

STARTING AN EROTIC JOURNAL

Above and beyond the benefits of responding to the SES, you'll gain even more if you keep an erotic journal, which is simply a notebook in which you write memories, feelings, and impressions as they come to you. Once you establish the habit of using your journal it's easy to add details as your memory becomes more active—which it surely will. Starting a journal also allows you to reread your comments as your self-discovery deepens; perhaps later you will see them in a new light.

Please keep in mind that any writing you do in the SES or your journal must be *for your eyes only.* Your writing will touch on a host of possible topics you might want to discuss with someone when the time is right. Such intimate exchanges can be extremely useful and fulfilling. But by keeping the writing itself private, you will avoid subtle inhibitions that can cause you to hold back, perhaps without even realizing it.

EXPLORING FAVORITE FANTASIES

As you recall some of your memorable encounters, I also want you to start thinking about another dimension of eroticism with which some people are more familiar than others—the realm of the erotic imagination. Sexual fantasies take an infinite variety of forms, and each individual has his or her unique patterns and preferences.

Fantasies spring from the depths of your erotic mind and are invaluable sources of information, which is why I included questions about them in the SES. I'm well aware, however, that the complexity of sexual fantasies, along with the fact that imaginary erotic scenes often do not conform to our ideals or values, makes them more difficult to explore than real-life encounters.

Some people, women more than men, aren't aware of fantasizing at all. Others fantasize primarily about exciting events that actually occurred in the past, perhaps with a few embellishments. Another common form of fantasy is imagining sexual possibilities you hope might happen one day—as when you daydream about a sexy stranger whom you pass each day on the way to work. Keep in mind, however, that fantasies don't necessarily bear any relationship to real life. If you grant yourself the freedom to do so, you can enjoy, within the sanctuary of your mind, fantasy scenarios that you would never want to experience in reality. Many people are confused on this point, mostly because they haven't allowed themselves to experiment sufficiently with the infinite flexibility of fantasy.

At the most basic level, many people are confused about what, exactly, a fantasy is, so I included this statement in the SES:

> An erotic fantasy is an image, thought, or feeling within your mind that is sexually interesting to you. Some people think of fantasy as a sexual daydream. Maybe it turns you on just a little bit—so little that you hardly notice. Or maybe it turns you on a great deal. Sexual fantasies may or may not make your body become aroused.
>
> A fantasy can be triggered by something you actually see or hear (for example, an attractive person or an erotic picture or story) or it can just pop into your mind out of the blue. It's very common for people to have fantasies while they masturbate, but it's also common for people to fantasize while having sex with a partner or while doing virtually anything else.
>
> People have many different kinds of sexual fantasies, and some people say they have none at all. A fantasy may be a simple or elaborate story—perhaps based on a past experience, a hoped-for encounter, or a totally imaginary scene. I ask that you pay close attention to your fantasies and how

they help to turn you on—even if you think your fantasies are boring, silly, or uninteresting. Remember that a fantasy does not have to be a big production in order to be important.

Three of the SES's most important and provocative questions direct your attention to the imaginative creations of your erotic mind:

1. Imagine yourself really wanting to be sexually aroused but, for some reason, you're not. Based on everything you know about your sexuality, describe the fantasy that would be *the very most likely* to arouse you.
2. What are your ideas about what makes this fantasy so exciting?
3. Describe the "climax"—the most intense point of excitement—of this fantasy.

Although the SES asks you to focus on one favorite fantasy, feel free to recall as many as you like. You may have a wide variety of fantasies. If so, jot them down. The question about the climax of your fantasy is intended to help you identify the specific details that intensify your excitement.

If you're a person who doesn't fantasize, don't be concerned. As you read on, try to keep an open mind. You may not be accustomed to noticing the sexy images that flash through your mind. Like many people, you may not have detailed fantasies, just fragments of erotic thoughts that easily go unnoticed. Moreover, like so many of us, you may have been taught that it's wrong to have sexual daydreams. Be patient and self-accepting. Gradually, you'll become more familiar and comfortable with the way your erotic imagination works.

As you contemplate your fantasies, don't be surprised if you feel embarrassed. After all, you're zeroing in on extremely intimate material. Most of us will tell a friend details about an exciting encounter much more readily than we'll talk about what truly arouses us in our private thoughts. If you can open the door only slightly to this line of self-exploration, in time you'll discover how much your fantasies can teach you about your eroticism.

A WORD ABOUT PEAKS AND PROBLEMS

At one time or another most of us experience problems with our sexuality. In fact, one of your motivations for reading this book may be to understand or to resolve your own sexual concerns or those of someone you love. I bring this up now because you need to be aware that in the five chapters of Part I our goal is to unravel some of the mysteries of the erotic mind. We'll do that primarily by focusing on peak erotic experiences. This is a different approach from the one most books about sex follow.

Like many of my clients, you might find it difficult to set aside your concerns for a while and see what you can learn from your own and others' peak turn-ons. I'm certainly not suggesting that you ignore your problems or pretend you don't have them; that would be counterproductive, not to mention impossible. It's best if you remain aware of your problems in the back of your mind while you focus on your potentials. I can assure you that your patience will pay off. By the time you reach Part II, "Troublesome Turn-ons," you will have developed the insights necessary for understanding and resolving a variety of prevalent erotic problems from a whole new perspective.

EROTIC MEMORABILITY

Now you are ready to begin examining your peak erotic experiences. Think of them using two seemingly mismatched metaphors. Peak turn-ons are precious jewels. To fully appreciate their glittering facets, it is necessary to gaze at them from different angles. Yet peak experiences are also onionlike. As each layer is peeled away you uncover additional information not visible on the surface.

All reminiscences, sexual or otherwise, are shaped by the way your memory sorts and stores information.[5] Vivid recollections are enlivened by what I call *memorability factors*—characteristics that make any event stand out. Typically, memorability factors are the most readily identifiable aspects of an experience, the special particulars and circumstances that first come to mind when you think about it.

Consider the memorable encounters you're recalled thus far. Which specific situational details contributed to your arousal? We're not looking for penetrating analysis here, just the obvious facts that helped make the experience notable. Record your observations in your journal.

Next note the specific details that stand out in your favorite fantasies. You'll probably notice important differences between the details that excite you in real life compared to those that are exciting in fantasy. In actual encounters, for instance, arousing details often appear to be strokes of luck, opportunities you happily seize even though they aren't necessarily of your own creation. By contrast, in fantasy you select and control all the exciting events. In the realm of the erotic imagination you are the creator as well as the director, with the power to make everything turn out exactly as you wish.

These memorability factors appear in The Group's stories more frequently than any others:

Firsts and surprises

Idyllic situations or partners

Extensions and restrictions of time

Learning about factors that stand out for The Group may call your attention to similar circumstances that contribute to your own arousal. But it's quite possible that other memorability factors are more significant to you. After all, few things in life are more personal than sexual excitation.

FIRSTS AND SURPRISES

The first time you have a significant experience you're more likely to remember it because the characteristic of "firstness" stands out against the background of the familiar. Similarly, when your expectations and routines are shaken by a surprise, you also tend to take notice. These principles of memorability help explain why nearly a third of The Group's peak encounters include at least one first or surprise.

Over the years I've consistently observed that men recall their first sexual encounters fondly as welcome rites of passage, even if the details of what happened weren't that terrific. Women, however, don't always welcome sexual initiation so wholeheartedly. Often a woman's first encounter "just happens" after a few drinks or in response to pressure from her partner or peers. Even when the encounter is desired and basically positive, her enthusiasm may be tempered by a sense of loss.

Despite these differences in how men and women feel about sexual initiation, among The Group neither gender is inclined to report its first sexual encounters as peaks, with the exception of a few people whose sexual initiation coincided with falling in love. Other firsts, however—such as initial encounters with new partners, experiments with new sexual activities, or encounters in new settings—are mentioned regularly.

Women in The Group often recall the first time that sex was truly satisfying for them or the first time they experienced an orgasm with a partner. Darlene, a thirty-nine-year-old respiratory therapist, tells what happened to her only a few years ago:

> I had fallen in love with my best friend. He wasn't physically "my type," but I thought about him constantly. I wasn't sure how he felt about me so I tried seducing him with the most romantic music I could find and asked him to read a poem I had written about him.
>
> All the while I was pessimistic. I thought to myself, "Even if the seduction is successful, the sex probably won't be so great." Before that night, I only had orgasms with masturbation. I always fantasized I could come during intercourse, but didn't believe I really could.
>
> I was so overcome with happiness when he said he was falling for me too. Soon we were making love. Usually after fifteen or twenty minutes of some guy pumping up and down, I would fake an orgasm just to get it over with. But after only a few strokes I was coming! Even though his penis was smaller than some other men I've been with, he fit me perfectly, hitting all the right spots.
>
> I had three orgasms that night. He just naturally did things that drove me crazy. They say you can't expect your partner to read your mind but to this day he always knows how to make me come.

Darlene's breakthrough demonstrates how firsts typically reveal important new facts about who or what turns us on. Like Darlene's, many of The Group's enjoyable firsts are especially memorable because they contain an element of surprise. Megan, a young college student, surprises herself—and her grateful partner—when she opens the door, nervously at first, to a new source of pleasure:

> My boyfriend had often asked me if I would give him a blow job. But I was scared because we had been seeing each other only five months—long enough for us to have sex, but I wasn't completely comfortable with him yet. And besides, I just wasn't sure it was the right thing to do or if I could do it right.
>
> I don't know why, but all of a sudden I had this thirst for knowledge that *had* to be quenched. We were in my bedroom when I surrendered to my curiosity. He was very loving and patient as he guided me. He showed me exactly where it felt best, almost like an anatomy class. He let me experiment on him with my lips and tongue. His moans told me I was on the right track. I felt adventurous and—it was weird—kind of in control. The combination of spontaneity and nervousness created an arousal all its own, especially when I saw my boyfriend's face as he came.

Isn't it striking how Megan's reluctance dissolves into a burst of curiosity and desire? But notice the necessary conditions: (1) *she* decides when the time is right, and (2) her patient partner helps her establish safety so that (3) she can transform her nervousness into an arousal enhancer. As a result, she moves from uncertainty to confidence and is validated by her boyfriend's responsiveness.

Another type of surprise occurs when familiar people behave in unfamiliar ways. Manuel tells of an unexpected conclusion to what appeared to be a colorless day:

> I was hanging out with my girlfriend. Nothing much was happening—a little shopping, visiting a couple of friends. We were bored as shit to put it bluntly. Finally we gave up and went back to her place. Another Saturday night watching TV.
>
> I needed a hot shower to give me a lift. Just as I was

> soaping up I heard the bathroom door squeak. Suddenly her hand came through the shower curtain, she grabbed the soap out of my hand and jumped in with me. We'd been together for almost three years and she'd never acted like *this* before. Her aggressiveness blew me away—but I liked it.
>
> She got on her knees and soaped up my dick. It was as if she had studied how I masturbate (even though she never saw me do it) and added her own special touches. Was this my Angela? She brought me to a peak of ecstasy like never before. My orgasm was an explosion. We continued our adventures in bed for a couple more hours. Boredom was a thing of the past.

For Manuel, the foreground (his intense excitement) stands out in bold relief against a background of boredom and low expectations. I predict that as you construct your own list of peak turn-ons, you'll find at least one that contains a similar contrast. In fact, The Group *only* brings up the topic of expectations in their tales of peak sex when those expectations are being joyously shattered by a welcome surprise.

In the coming chapters we will often be reminded that peak encounters share many features with memorable fantasies. But here's a fascinating exception: while firsts and surprises obviously help make peak encounters memorable, they seem to be of little significance in fantasy. Only 5 percent of The Group's favorite fantasies involve anything unexpected or surprising, and *none* of them contains a first of any kind.

What shall we conclude from this? Our best encounters often take us to places we've never been. Favorite fantasies, on the other hand, cover familiar territory. Through repeated experimentation we refine them so that they express, in the shorthand of imagery, the essential elements needed for arousal. This amazing capacity of erotic fantasy to hold our interest in spite of repetition is something we'll return to many times, especially in Chapter 5, "Your Core Erotic Theme."

IDYLLIC SITUATIONS AND PARTNERS

Tropical islands, warm breezes at sunset, toasts to love over candlelight, meaningful glances from a beautiful stranger—these and countless other idyllic situations are associated with romantic fantasy and fiction. Nevertheless, 21 percent of The Group's real-life encounters stand out precisely because they involve unusually fantasylike settings or ideal partners. Not surprisingly, many of these encounters take place on vacation. Far away from familiar surroundings and distractions, it's easier to set aside inhibitions, to take a risk, to initiate or renew a romance.

Such encounters can also unfold much closer to home, as in the case of Trevor, a gay man in his late thirties:

> His name was Eric and he was extremely attractive to me, with a firm, slightly developed body. He was the absolute best hugger. My body came alive when he wrapped me up in his arms. I especially enjoyed kissing him, his lips so soft as he kissed me in return. I remember gazing deep into his eyes while he fucked me as I sat on top of him. Our motions were in perfect harmony. It was easy handling him inside me.
>
> I was totally amazed by it all and kept staring at him and his beautiful body, wondering if it was all a dream. The thought of my good luck, together with the sight of him, and the feeling of his dick inside of me—I've never been more aroused. His movements and thrusts when he came gave me an orgasm without any stimulation of my cock. I couldn't believe it since I've always needed my dick jerked off to come.
>
> I never saw Eric again. Nor have I ever felt so responsive since, though I've attempted to recreate the feeling with other men. I often wonder what made that evening so unique. It was truly magical.

In addition to its idyllic features, this story also has a poignant quality. While Eric clearly personifies Trevor's ideal lover—with his beautiful body and ability to express tenderness and affection—he also represents a taste of perfection that seems to be a once-only bit of luck, never to be repeated. A similar hint of wistful longing can often be perceived in tales of idyllic encounters. They have a dreamlike, otherworldly quality that, by definition, is quite rare.

In fantasy, however, it's easy to create at will an ambiance of perfection, which is what well over one-third of The Group does in their favorite fantasies. Arlene's fantasy captures the feeling perfectly:

> I'm in the mountains and have floated out to a rock in the middle of an isolated lake. I am lying in the sun, soaking up the warmth, with no clothes on and none with me. Suddenly I'm aware of a handsome man in his early to mid-thirties, at least six feet tall, slender, muscular, with dark hair. His body is in great shape. He's beautifully tanned with soft lips, talkative eyes, and large hands. He is naked too.
>
> He swims over to my rock and climbs up. Slowly, passionately, he kisses me and then licks every part of my body, one by one. I can barely stand it when his tongue wiggles its way up my thighs to my vagina where he meticulously traces the shape of each lip, circling the opening and then kissing my clit until I'm writhing in ecstasy. He lies down beside me and soon we make joyous love. First he is on top of me, then I'm on top of him. We are free, incredibly sensuous and tender. Afterward I quietly swim off as he sleeps. I glance back for one last look at his moist body glistening in the sun.

Arlene's fantasy is quintessential "high romance," far more common among women than men. I've often noticed that women frequently make a point of sketching out dreamy settings, sometimes concentrating their attention on the environment as much as or more than on sexual specifics.

When a man describes idyllic fantasies, with rare exceptions he emphasizes the perfection of his partner's body, usually with little or no interest in the environment, except as a setup for great sex. Like Arlene, Luke enthusiastically dwells on the exquisite beauty of his ideal fantasy lover. The ambiance, though, is notably different:

> I'm at home in my apartment watching TV in my gym shorts when the doorbell rings. I can't believe my eyes when a gorgeous fox is standing there. She's my new neighbor and just stopped by to say hello. I eagerly invite her in. She's drop-dead gorgeous in a silky nightgown that reveals every curve. Her waist is narrow, her hips wide and shapely. I watch her ass sway as she walks to my sofa. Long, auburn hair swoops down, partially covering the milky skin of her cleavage. I

can't help staring at nipples which show clearly through her gown.

I'm fumbling for words when she slides closer to me and plays with my chest and stomach with her long fingernails. I pull her closer still as she grabs my dick. I feel a shiver go through her body.

I invite her to the bedroom and lift the gown over her head, revealing an even more incredible body than I expected. She rips off my shorts and we fall into bed, fucking with uncontrollable abandon. She loves it when I plunge into her juicy pussy. Before I know it she's coming wildly, screaming. Her reaction turns me on so much that I thrust faster and faster, coming, coming until I collapse on top of her, spent. After I calm down, I watch her slip into her gown, shake out her hair, and walk toward the door. "See ya later, neighbor," she whispers as the door closes behind her.

Luke and Arlene both enjoy the surprise of a perfect stranger. But whereas her fantasy is steeped in romance, his is animated by unfettered lust. She surrounds herself with mountains, a lake, and a sun-drenched rock for a bed. His perfect lover simply arrives at the front door. Also absent from Luke's fantasy is the gentle tenderness that is so crucial to Arlene. Although such differences between men and women are by no means universal, they are unmistakable.

EXTENSIONS AND RESTRICTIONS OF TIME

If you've ever found yourself glancing at the clock in a sexual situation, things probably weren't going so well. The ticking of the clock as time marches forward is an apt metaphor for the mundane repetitions that occupy so much of our lives. No wonder we normally lose all consciousness of time during fulfilling sex, except perhaps for a fleeting wish that this moment would never end. In some situations, however, an awareness of time actually contributes to the enjoyment. More than one-fifth of The Group's descriptions of peak encounters, and almost as many of their favorite fantasies, contain references like these:

"This went on for hours."

"We never left the cabin all weekend."

"We made love all night."

"It seemed to last forever."

"We took the time to savor every moment."

Time lavishly devoted to an erotic adventure is a testament to its importance. Only highly significant activities command such attention. Lydia sings the praises of extended lovemaking when describing a particularly memorable encounter with Josie, her lesbian lover:

> Josie and I live such busy lives that we usually squeeze in some routine sex on the weekends. One Sunday I fully expected a typically brief lovemaking session. I knew things were going to be different when Josie spent a good twenty minutes lovingly licking *each* of my breasts. My excitement built so gradually that I went into an altered state of consciousness—as if each movement was acted out in slow motion.
>
> Josie seemed to go into a similar state when I went down on her. Normally I get bored after a while, but that day I savored her cunt as if it were my last meal. And speaking of meals, we fixed a delicious lunch of cheese and fruit and crawled back into bed to eat it sensuously. We held and kissed and rubbed and tickled and laughed all afternoon, at one point spontaneously drifting off into a semislumber in each other's arms.
>
> We both had many orgasms that day, but I know mine were different than usual. They were slow pulsations or undulations rather than the rapid contractions I'm used to. We forgot about everything, focusing all of our energies on each other. Spectacular!

In other instances, time is a memorability factor for the opposite reason: because there's so little of it. Stolen moments with a secret lover, a hurried outdoor tryst, a passionate embrace in an elevator, a "quickie" before running off to work—all stand out because time is scarce. A desire so intense that it *demands* expression, even when there is insufficient time for it, demonstrates its compelling urgency. Norman recalls with enthusiasm one evening when he and his girlfriend were rushing to get ready for a concert:

> Tammy and I often disagree about who should initiate sex, when, how often, and how long it should last. Sometimes it can be such a pain in the butt I'd rather avoid the whole thing. But there have been several times when all that crap goes out the window. This usually happens when we're running late for something. Knowing that nothing will come of it I find it easier to be passionate, like one night when Tammy was dressing for the symphony.
>
> I rubbed her shoulders and she tried to push me away. But I wouldn't quit. I enjoyed turning her on even though she whined, "Norrrrman, we're going to be late." Next thing you know I was kissing her neck and reaching in her panties. All of a sudden she became like an animal. She grabbed me and kissed me deep and hard while I rubbed her clit and brought her to an orgasm in a minute or two—*much* faster than usual. Just a few strokes of my cock and I came too.
>
> Then we went flying out the door, laughing like lunatics. At the concert she told me there was lipstick smeared on my face. We couldn't stop laughing. Now why can't it be like this all the time?

Dr. Maslow noticed a curious phenomenon, difficult to explain or even describe, in his research on all kinds of peak experiences: pleasurable distortions of time and space. He made this observation:

> Not only does time pass in their ecstasies with a frightening rapidity so that a day may pass as if it were a minute, but also a minute so intensely lived may feel like a day or a year. It's as if they had, in a way, some place in another world in which time simultaneously stood still and moved with great rapidity.[6]

Although this sounds rather "cosmic," if you've ever had any kind of peak experience, you probably sense what Maslow's getting at.

PRACTICAL USES OF EROTIC MEMORABILITY

Just because peak experiences can't be ordered on demand, you need not wait passively. Knowledge of which memorability factors have contributed to your arousal in the past can help you cultivate conditions for more fulfilling sex now and in the future.

For instance, passionate lovers who appreciate surprises become adept at deliberately breaking their sexual routines with playful experimentation. Creating unexpected turn-ons isn't simply a matter of behaving less predictably—although this certainly helps. It also requires the ability to increase your capacity for being surprised. Zestful lovers allow themselves to be caught off-guard. On the other hand, I've often noticed how bored lovers develop an uncanny ability to miss opportunities for surprise, usually because they stop paying attention.

Too many people also assume that they can only wait and hope for idyllic situations or partners to bring special excitement to their lives. Passionate lovers discover that lucky moments happen more frequently to those who consciously devise the necessary conditions. And far more than we realize, the seemingly magical appearance of an ideal partner is a mixture of happenstance and a heightened readiness on the part of the beholder to perceive beauty. Likewise, those who have luxuriated in extended lovemaking or found a special charge in quick sex can learn to request and seize opportunities for similar satisfactions that might easily slip by.

PARTNERS IN ECSTASY

In virtually every peak encounter or fantasy, the partner—more than one sometimes—plays a major role. Who *are* these people? Scan your peak encounters and take note of the kinds of relationships you've had with the partners in them. Then notice the partners you've selected for your favorite fantasies. Are they similar to or different from your real-life partners?

If you've already looked over the SES you know that after each peak experience you're asked to categorize the relationship between you and your partner as:

Anonymous

Acquaintance

Boyfriend/girlfriend

Primary relationship or spouse

Multiple partners

In the original version of the SES I didn't realize I needed to ask this question because I assumed the information would be obvious from the stories. So you can imagine how surprised I was to discover that 16 percent of The Group's peak encounters and 21 percent of their favorite fantasies contain no information whatsoever about the relationships they have with their partners! During certain moments of lusty sex, apparently the type of relationship doesn't matter—only the quality and intensity of the interaction. Luckily, in approximately 80 percent of their stories The Group *does* mention what kind of relationships they had with their partners.

As you can see from the stories you've read so far, The Group's partners in ecstasy range from total strangers to committed spouses. Closer analysis reveals that both the gender and the sexual orientation of the storytellers affect the kinds of partners with whom they are likely to have peak sex. But the most important consideration of all is whether the partner (or partners) is a real person in an actual encounter or a fantasy partner.

REAL-LIFE PARTNERS

Almost 60 percent of The Group's peak encounters costar a partner with whom they have at least some degree of involvement. "Acquaintance" and "boyfriend/girlfriend" are the most frequently mentioned categories, followed by "primary relationship or spouse." Even though men are slightly more likely than women to be currently in a primary relationship, their peak encounters are less likely than women's to involve a primary partner or spouse. This does not necessarily mean that the men go outside their primary relationships for peak sex, although some do. More common, they recall peak encounters that occurred before they met their current partners.

A number of men and women report that their most memorable encounters occurred with people who later became their primary partners. During the early stages of such involvements the intensity of infatuation is normally at its highest. But there's a significant gender difference here. Women are almost twice as likely as men to mention that they feel romantically involved or in love with their partners in ecstasy (25 percent and 13 percent respectively).

As I report research findings from the SES in the coming pages and chapters, you'll notice that lesbians almost always have similar experiences as women in general, but typically with greater frequency. Here is the first instance of this phenomenon: lesbians are the most likely of all subgroups to speak of love and romance, doing so in 36 percent of their encounters as compared with 26 percent for straight women. Straight men mention loving their partners in 11 percent of their peak encounters, while almost one-fifth of the gay and bisexual men enthusiastically express love. For reasons I can't explain, none of the bisexual women mentions being in love with her partner.

Almost one-quarter of The Group's peak encounters involve complete strangers or people they've just met. But here we see another great gender difference: men are almost twice as likely as women to describe very casual or anonymous encounters (30 percent and 16 percent respectively). Reflecting this trend, the most likely of all subgroups to have anonymous encounters are bisexual men (50 percent) and gay men (47 percent). At the other extreme, only 1 percent of the lesbians describe memorable sex with strangers or near-strangers.

Seven percent of The Group's peak encounters involve two or more partners, mostly "three-ways." Men are much more likely to report multiple partners (13 percent) than women (3 percent), a tendency that is particularly evident among gay men and lesbians. Whereas 18 percent of gay men's encounters involve multiple partners, none of the lesbians' do. Matthew, a gay college administrator, describes a particularly successful three-way:

> I saw two male lovers at a nude beach and enjoyed attracting their attention by getting a hard-on and letting them see it. Eventually one of them walked toward me at water's edge. He had a thick, beautifully proportioned cock. Even when soft it swayed heavily and slapped from thigh to thigh as he walked. Soon after we greeted he invited me to hang out with them. All afternoon we joked and flirted. They were both extremely handsome.
>
> I was thrilled when they invited me to dinner. At their place we all took turns showering. When one of them was in the shower, the other would play with and suck my cock. They had been together five years and claimed this was their first three-way. Both turned out to be incredibly sexy. I could see into their bedroom from the kitchen as each dried off and stroked their hard dicks.
>
> After dinner one of them got down on his knees in front of me and said to his lover, "Look what I've got for you," as he pulled down my shorts and began to suck me. By now the other one was feverishly jerking off.
>
> When we made it to the bedroom I was thrilled to watch them perform sixty-nine on each other—*very* hot. I ended up fucking one lover as I was fucked by the other. Both men were warm, gentle, with great senses of humor. One also sang beautifully. They obviously loved each other. Our three-way never became competitive or tense—not even for a moment. Both were eager to please me, and I them. I can't imagine how it could have been better.

Bisexual women are the most likely of all to report peak encounters with multiple partners (25 percent), and bisexual men run a close second (20 percent). Such encounters offer obvious advantages for bisexuals, as is apparent in this steamy tale of group sex at a hot tub party told by Ginny, a college instructor in her early forties:

My boyfriend, Rob, had been coaxing me for a long time to attend one of those "swinging" parties where it's considered okay to sample sex with different people. At first I was reluctant because I feared the scene would be much too sleazy for my tastes. Being a great salesman, Rob emphasized the fact that my previous lover was a woman and that I might enjoy watching both sexes getting it on.

Eventually I agreed and much to my surprise it wasn't seedy at all. The party was at a beautiful home with a large pool and hot tub. The people were intelligent and friendly. Other than the nude sunbathing it was no different than any other party. Rob was right, I certainly did enjoy gazing at the wonderful collection of bodies.

After the sun went down the atmosphere became more sexual. Some people went inside, apparently to have sex in one of the many bedrooms. Rob and I joined the group in the hot tub. As we loosened up, people began massaging each other. One woman was sitting on the edge when another woman went down on her. I was so spellbound by that sight that I didn't even realize that Rob was getting a hand job—until I heard his familiar groans. When I turned around his face told me he was about to come. He did and several people cheered. I was a little jealous but incredibly turned on.

Over the next few hours, we put the lounge cushions on the deck around the tub so we had a choice of comfortable positions. I especially enjoyed being touched by Rob and a beautiful young woman at the same time. It was all so friendly and warm and fun—not at all like my idea of an orgy. Although he denies it, I think Rob was more jealous than I because he never mentions trying it again.

Matthew and Ginny are among the relatively few members of The Group who describe group sex encounters as their most memorable. These encounters are fraught with potential problems and are much more difficult to arrange than the more common one-on-one variety. Typically one partner is more interested in a multiple-partner adventure, which can result in arguments. Once partners do agree on a three-way or group sex scene, somebody often feels left out or jealous. And what are the odds of three or more people having compatible sexual desires? All of these complications vanish, however, in the realm of the erotic imagination.

FANTASY PARTNERS

One of the most striking differences between The Group's favorite fantasies and their actual encounters is their partners. Fantasies involving multiple partners are the most popular of all among The Group as a whole. Just as they are the most likely to have multiple partners in real life, bisexuals are the most drawn to fantasies of multiple partners (60 percent of bisexual men and 75 percent of bisexual women). But more than a quarter of all women—even a third of the lesbians—say their favorite fantasies involve two or more partners. The same thing is true for 43 percent of the men.

One finding may surprise you as it did me. Whereas almost one-fifth of gay men enjoy multiple sex in their memorable encounters, they are the *least* likely subgroup to include multiple partners in their favorite fantasies (only 11 percent). The best explanation I can offer is that virtually any gay man who wants to try group sex or a three-way can find opportunities to do so, particularly in urban areas. These activities were especially widespread in the freewheeling days before the AIDS epidemic. It appears that real-life experiences with multiple partners reduce their allure in fantasy.

What is the meaning of the special appeal that multiple-partner fantasies hold for so many men and women? The ubiquitous imagery of two eager women in male pornography undoubtedly reflects and reinforces men's interest in three-ways. But what about women? Their most popular form of erotica—the romance novel—virtually never includes multiple partners. With rare exceptions, such as when three people fall in love with one another, multiple partners do not easily fit the romantic ideal. Fantasies involving more than one partner typically have a purely lustful quality.

Many factors contribute to the popularity of multiple partners—especially three-ways—among The Group's fantasies. The fantasizer is virtually always the focal point of such scenarios. The role of both partners is to respond to every whim of the fantasizer and in doing so to affirm his or her irresistability. In addition, the fantasizer is always in control, whether he or she chooses to dominate, to submit, or prefers to watch the partners put on a show as they have sex with each other.

I believe the most important attraction of three-ways is their ability to amplify whichever characteristics turn the fantasizer on. Typically, both partners are of the same gender and thus provide a double dose of maleness or femaleness. Consequently, straight women and gay men usually imagine two or more men, whereas straight men and lesbians gravitate toward two women. Not surprisingly, bisexuals sometimes enjoy the presence of both genders, but many prefer to take advantage of the amplification effect by fantasizing about two men or two women, depending on their inclination at the moment.

Second only to the popularity of multiple partners in favorite fantasies are very casual or anonymous partners. Among most of the subgroups, regardless of gender, 20 to 24 percent of their favorite fantasies involve sexy strangers or casual, chance meetings. Bisexual men have the most fantasies of anonymous sex (40 percent) and lesbians have the fewest (17 percent). In real-life encounters most women want some link between sex and feelings of emotional connection, as compared with a significant number of men who do not necessarily require or even want such a connection. However, this distinction almost completely disappears in fantasy. It is a dramatic reminder that in the realm of the erotic imagination we are frequently exempt from the values and preferences that guide our actual behavior.

In only 12 percent of cases does The Group select fantasy partners with whom they have any real involvement beyond their fantasies, whether as dates, boyfriends or girlfriends, or primary partners. Women, however, are more likely than men to fantasize about partners with whom they're involved (14 percent and 9 percent respectively). An even greater gender difference appears in regard to being infatuated or in love with their fantasy partners. Women mention feelings of love more than three times more frequently than men (14 percent and 4 percent respectively). And once again, lesbians are the most likely (17 percent) to mention loving their fantasy partners.

This brief overview of the kinds and numbers of partners involved in peak encounters and fantasies underscores how frequently highly arousing experiences deviate from the norms and ideals with which most of us are raised. How often have we read or been told that sex is best with a loving partner? Yet The Group's turn-ons remind us that memorable and thoroughly enjoyable sex occurs in the context of all sorts of relationships, sometimes with no emotional connection at all. Particularly in the realm of fantasy, the erotic mind claims for itself a wide zone of freedom from social conventions.

PREPARING TO DELVE DEEPER

You have launched your exploration of eroticism by focusing on the aspects of peak turn-ons that are the easiest to observe: (1) memorability factors that help make these experiences stand out, and (2) the partners who serve as objects of your erotic fascination. However, I'm sure you've already sensed other aspects of your memorable turn-ons that aren't so obvious. The deeper, more complex dimensions of peak erotic events will occupy our attention in the coming chapters.

If you have not yet started writing, why not begin now by responding to the SES or starting an erotic journal or both? I guarantee you'll get much more out of this book if you use these tools to personalize your observations. If you become a participating observer rather than a detached one, your discoveries will be much more likely to enrich you.

LETTING GO OF JUDGMENTS

In the introduction I emphasized how the erotic mind reveals its secrets only to accepting observers. This idea may have sounded reasonable enough, although perhaps a bit abstract. But now that you're actually looking at some of the most intimate expressions of eroticism—both your own and others'—you probably have a more concrete sense of how judgmental thoughts can inhibit you. As you venture further, you're even more likely to encounter material that is difficult to understand, and thus easy to criticize.

Think back over the peak encounters and fantasies you have remembered so far. Have you noticed anything that you felt inclined to judge? If you're keeping an erotic journal, start a new section entitled "Judgments and Criticisms." Each time you notice a critical thought toward one of your turn-ons, simply note your reaction. Avoid the circular trap of judging yourself for being judgmental. As your awareness of judgments increases, so too will your ability to set them aside.

Try the same experiment as you read The Group's stories. You may have found some of them a bit kinky. As you read on you will encounter even more unconventional turn-ons, in other people and perhaps in yourself. Pretending to be open-minded is of little use. It's much better to acknowledge whatever you find disturbing or difficult to accept. Jot down how you feel in your journal. When you come across a story that stimulates a particularly strong reaction, ask yourself how you might feel about the story if you approached it with an attitude of neutral curiosity. When you adopt such an attitude, if only for a moment, you'll notice something quite remarkable: behaviors and fantasies that initially seemed weird or unacceptable become increasingly comprehensible.

KEEPING PEAKS IN PERSPECTIVE

Almost three-quarters of The Group say their peak turn-ons are much more or dramatically more arousing than their typical sexual experiences. This discrepancy raises important questions: What is the relationship between peak arousal and the regular, everyday kind? Does studying particularly exciting sex help us produce additional satisfying experiences, or is there a danger that we might end up feeling disappointed with simpler, less earthshaking pleasures?

How unfortunate if we use the perfection of our best experiences to devalue more mundane sex. Sex therapists regularly see clients who have converted moments of special pleasure into sources of disappointment and frustration by using them to create higher standards and, in turn, greater pressures to perform. Tragically, they have turned the beauty of their peaks into painful reminders of their inadequacy.

What is the alternative? Peak turn-ons bestow their gifts most generously when each is recognized as one-of-a-kind. All peak experiences spring from total involvement in the moment, which is lost if you split your attention by comparing one moment to another. However, when you savor each magical memory on its own terms, your recollections help you to become more fully available for a wider variety peak erotic experiences.

2

THE EROTIC EQUATION

Flames of passion are fueled by a mixture of attractions and obstacles to overcome.

In the latter part of the 1970s my professional interests and personal struggles coincided as never before. On the personal side I had just extricated myself from the most painful yet sexually exhilarating relationship of my life. At one moment we would be lost in passion. Then, without warning, my lover would vanish, apparently overwhelmed by our closeness. For years I had come back for more until, devastated and humiliated, I eventually broke it off for good. As I mourned my loss, I wondered whether—if I ever let myself fall in love again—I was destined to repeat the same drama.

In my professional life, especially my studies of eroticism, I was also at a turning point. For years I had enthusiastically adhered to the principles of modern sex therapy launched by Masters and Johnson less than a decade before. They had made thousands of therapists realize that asking clients to try structured experiments at home could often help them work through even long-standing sexual problems more effectively than traditional therapies. This approach had a worthy goal: to reduce or eliminate impediments—such as performance anxieties, unrealistic expectations, or guilt—that block "natural" sexual responsiveness.

It was becoming increasingly clear to me that sex therapy's focus on the removal of impediments to sexual desire and arousal was far too narrow. Whereas modern sex therapy is anchored in the neat-and-clean model of sexual interaction that views barriers and inhibitions as unnecessary and unwelcome troublemakers, I was finding it impossible to ignore the fact that barriers seem to turn people *on* at least as often as they turn them *off*.

Then, in preparation for a talk on sexual orientation, I was reading *The Homosexual Matrix*, an exceptional book by psychologist C. A. Tripp. In his chapter entitled "The Origins of Heterosexuality," two sentences jumped out at me:

> A person's sexual motivation is seldom aroused and is never rewarding unless something in the partner or in the situation itself is viewed as resistant to it. This resistance may be in the form of the partner's hesitance, the disapproval of outsiders, or any other impediment to easy access.[1]

Not only was Dr. Tripp talking about me and my fizzled romance, but he was also addressing exactly the kinds of contradictions haunting me in my work. As I read on, my self-preoccupation gradually gave way to the realization that my own torturous struggle reflected a larger human drama:

> It is apparent that an erotic attitude does not develop toward a fully accessible partner (even one who's wonderfully complementary), but is aroused like a cannibal's appetite when a desired but somehow remote partner cannot "be had" by other means. As anyone can see, sexual motives are especially stimulated in a person who feels an urgent need or intense admiration for the qualities he sees in a partner and wants to "import." Certainly there is much in sex that has to do with wanting, taking, conquering, or otherwise possessing a partner, sometimes one who has as little as a single highly desired quality.

I had to admit the obvious truth: my ex-lover's unavailability had been a key ingredient of the overwhelming intensity that had held me in its grip. Dr. Tripp's resistance principle provided me with a new way to examine and understand my patterns of desire and arousal—a perspective I have used ever since, in both my personal and professional life.

Over the years I've returned again and again to Dr. Tripp's concept, completely blending it into my ways of thinking about eroticism. It became my own when I discovered that the resistance principle was easier to grasp and use when translated into a simple formula I call the *erotic equation,* which summarizes a reality we've all experienced:

ATTRACTION + OBSTACLES = EXCITEMENT

Whenever I tell someone about the equation, whether I'm talking with a client in psychotherapy or addressing a group of therapists at a seminar, I'm struck by how frequently people have the same contradictory responses I did when I first read Dr. Tripp. Everyone knows it's human nature to want what we can't easily have. Some experiences, it seems, are so universal as to be virtually invisible. When the erotic equation restates for people something they already know, they typically react as if a light has pierced the darkness. Suddenly they realize why they're not always attracted to the "right" people, or why the unavailable ones are often the most fascinating.

This is the heart of the matter: although sexual desire and arousal can be stimulated by all sorts of people and situations, your most passionate responses spring from the interaction of two competing forces. First, an attraction pulls you toward the object of your desire. Perhaps part or all of that person's body closely matches your ideal. If you're *really* interested, you also perceive, accurately or not, certain personality traits you want to take in—as Dr. Tripp says, to import. And you want the desired one to see qualities in you that are worthy of exporting to him or her. Already the seeds are sown for your attraction to become truly dynamic.

Suddenly or gradually, your fascination comes up against one or more obstacles to overcome—the second requirement for a truly compelling erotic response. Perhaps it's unclear if your interest is reciprocated. Maybe the person is unavailable or somehow inappropriate to pursue. If your attraction touches a romantic chord, the risks of being hurt may loom up, urging you to retreat. Or you may simply be passing strangers, communicating with your eyes the thrill of something that can never be.

The erotic equation shows us why peak eroticism is rarely tidy, static, or predictable. It helps us fully grasp what we have always known: the erotic experience, by its very nature, is shaped by the push-pull of opposing forces and is therefore energetic, interactive, and potentially dangerous. We are the most intensely excited when we are a little off-balance, uncertain, poised on the perilous edge between ecstasy and disaster.

The idea that our most erotic moments are born of conflict is not new. Freud certainly recognized it, though he conceptualized it differently. He believed that an eternal tension exists between the primitive, sexual, animalistic id and the overcivilized super-ego. Freud brought a radical message to his Victorian contemporaries: though you may do your best to suppress sexual urges in yourself or your children, they're not going to disappear.

Like Freudian psychology, the erotic equation describes the interplay of impulse and restriction. But I've always thought that Freud had a dark and pessimistic view of human nature, perhaps because his lifework chronicled the ravages of sexual repression. In any case, he concluded that even reasonably well-adjusted adults were doomed to frustration because the requirements of civilization *must* prevail over our unruly impulses.[2] He didn't seem to realize that we humans want some restrictions to push against. We just don't want them to be so harsh that they choke off our sexuality entirely.

MYSTERIES OF ATTRACTION

In a world populated with an endless array of faces, bodies, and personalities, why are we sexually drawn to some, repelled by others, and indifferent to the great majority? Sexual attraction is one of life's great mysteries. People often act as if this mystery were a fragile one. They fear that looking too closely at their attractions might dampen or destroy them.

If you feel a similar reluctance to see all there is to see, let me offer you some assurances. I've talked in depth with hundreds of people who risked a closer look, and I've noticed a consistent pattern. Examining an attraction only disrupts or diminishes it if something about the attraction is detrimental to the person. In such cases, which we'll discuss in detail in Part II, looking at the dynamics of the attraction may set off a self-protective alarm as its potentially destructive features become more apparent.

Most people find that exploring their attractions deepens and enriches them. All of life's truly great mysteries share this in common: the more you see of their inner workings, the greater is your awe. If you approach your attractions with respect, they will reveal some of their secrets to you. Although you will never figure them out entirely, even small insights can enlarge the arena of your conscious choice—which is always empowering.

There are two primary types of attraction: lusty and romantic. Each springs from distinct motives and generates different kinds of passions. Those who aspire to a healthy erotic life must develop a comfortable relationship with both types of attractions, for each is part of our humanity.

LUSTY ATTRACTIONS

Dictionary definitions of lust mirror our mixed feelings about it, running the gamut from surprisingly positive to strongly negative. At one pole lust is simply pleasurable delight in our sensual appetites. It can also connote a strong enthusiasm, as in the phrase "lust for life." Most people see this kind of lust as admirable. Similarly, *Webster's* defines "lusty" as "vigorous, robust, and hearty"—nothing negative at all. At the opposite extreme, lust is defined as unrestrained, wanton surrender to carnal urges. From this point of view, a lustful person is often considered lascivious, lecherous, unsavory, and a potential menace.

Sexual lust is decidedly unpopular these days, firmly linked with disease, pregnant teenagers, sexual abuse, harassment, sexual addiction, and even lust murders. Given such unappealing associations, it may be difficult to think of it in a positive light. Thus the emphasis has shifted to relationships and monogamy.

There's been quite a change since the 1960s and 1970s, when sexual experimentation was widely celebrated. AIDS, of course, changed all that, but other factors also played a part. If you participated in that era's "sexual revolution" you got a pretty good look at lust in action and probably weren't completely comfortable with everything you saw. Many people, especially women, found that casual sex wasn't particularly satisfying.

Although lust has perhaps inevitably fallen into disfavor, we make a terrible mistake if we reject it completely. Our erotic health requires that we make room for lust, for it provides much of the zest that makes sex fun and self-affirming. Socially, it is also very important not to reject lust, no matter how relentless the antisexual clamoring may become. When lust falls victim to the forces of repression, its negative potentials increase dramatically.

At the heart of lusty attraction lies the desire for sexual excitation and orgasmic release, pure and simple. It can be profound, utterly meaningless, playful, loving, or hostile. In its most intense forms lust has an animalistic quality that can be exhilarating, frightening, or both. When you're feeling lusty your attention is focused primarily on whatever it is you want that produces and intensifies sensations of arousal, especially in the genitals.

LUST'S OBJECT

When you see someone who looks sexy, it seems as if that person is making you feel aroused, even though the source of arousal is your own mind and body. The sexy other is simply a stimulus and, at least to a degree, an object.

The nature of lust is to objectify, a reality that can be troublesome for many people. According to one popular line of thinking, to see a person as an object is to do him or her a grave injustice. People must always be regarded in their entirety, not merely "used" for selfish gratification. Focusing on just a part of someone for sexual kicks—voluptuous breasts, bulging biceps, or genitals, for instance—may even be considered a form of victimization.

Like most aspects of erotic life, lusty objectification isn't so simple. At its best it is an effective source of validation and approval. Having a desired partner perceive you as the object of desire can be flattering and exhilarating. Both men and women—although by no means all—crave opportunities to be responded to as sex objects, and more than a few bemoan the fact that it happens too rarely. And as a society we spend billions of dollars and untold hours trying to make ourselves attractive sexual objects.

To objectify is also to externalize, to recognize the desired one as the *other*—that is, to see clearly that he or she is outside oneself. This quality of otherness is absolutely essential for attraction. Not only is the object separated from the self, but that person is invested with sufficient value to make him or her worthy of pursuit.

One of the most beneficial features of lusty objectification is how it facilitates selective perceptions and idealizations. When you lust after someone, you naturally emphasize the qualities you find most appealing. Because lust focuses exclusively on turn-ons and screens out everything else, it's an extremely effective attraction booster. Even in an established relationship, in which you know and care for your partner as a whole person, look closely, and you'll probably notice how selectively attending to particular characteristics helps stir your passions when you're in a sexy mood.

Sonya, a thirty-eight-year-old member of The Group, describes how her fantasy life revolves around lusty objectification:

> I hardly ever have complete fantasy stories like the ones in books. When I want to get hot I just imagine a beautiful set of male buns. I love to scan my eyes from the wide, muscular shoulders, down the v-shaped back, to that sloping transition from back to butt. The very top of the crack thrills me, especially when I catch a glimpse of it at the beach when a hot guy is wearing a skimpy swim suit.
>
> A gorgeous set of buns calls out to be caressed by my eyes or fingers. I go nuts over ones with dimple indentations on the sides. At the moment I can't get enough of my boyfriend's buns. Like him, they're perfect! I'm always grab-

> bing him there which he seems to like. When I'm alone and horny I just think of him slowly peeling off his shorts while I watch from behind.

Whereas men have always readily described themselves as "tit men," "leg men," or the like, only recently have women, like Sonya, allowed themselves to admit to having a focused appreciation for specific physical attributes.

The strongest example of the objectifying quality of lust is a fetish, a superfocused erotic fascination with an inanimate object—something like underwear or shoes or garter belts—although the popular definition has gradually expanded to include a greater than usual fascination with a particular body part. The fetish object usually has some obvious link to sex, but not always. Yet almost everyone with a fetish knows the circumstances under which it developed and its erotic significance for them.

I once worked with a man who was very concerned about his obsession with raincoats, especially yellow plastic ones. His most intense orgasms occurred when he masturbated while wearing a raincoat, of which he had quite a collection. Although dismayed by his fetish, he had no trouble explaining it. As a boy he had received a gift of a little fire truck large enough for him to sit on and drive around the room. With it came firemen's gear, including a yellow raincoat. There were two things he especially liked about this toy: the tingling sensations he got in his groin when he rode the truck and the imagery of strong, brave firemen he conjured up in his mind as he playacted various rescue scenes.

Much later he came to realize that he was gay and that his fire truck sensations and fantasies offered him a compelling focal point for his fascination with men and masculinity. As years went by the masturbatory aspects of his raincoat rituals became more explicit and intense. With the addition of one more ingredient—an ongoing struggle to resist what he judged to be his "sickness"—he developed a full-blown fetish that continued to provide him with anxious pleasures throughout his adult life.

This story demonstrates how lust can become focused on a single object and the images that go with it. The fetish object becomes a kind of shorthand or, more accurately, an erotic cue that provides a pinpoint focus for arousal. Although most people don't attach all their sexual desire to a single object, normal lusty attractions nonetheless have an unmistakably fetishistic quality to them.

LUST AMONG MEN AND WOMEN

The differences between men's and women's attitudes toward lust are often debated. It was once widely believed that women had little if any interest in lust. We now know this isn't true. The expanding library of books of women's fantasies is a testament to the potential lustiness of the modern woman. There are, however, very real differences between men's and women's lust. The narrowing of focus that is a hallmark of lust operates in both sexes, although it is significantly more pronounced in men. I believe that a major reason for this difference is the penis—an instantaneous and unavoidable arousal feedback system.

Even as a children, boys are constantly learning about their turn-ons, even if they're not supposed to—even if they don't want to. A stiffening penis is extremely difficult to ignore. Later, when they learn to masturbate, most men discover an even more compelling link between their favorite fantasy images and the immediate responses of their genitals.

When a girl feels turned on, her genital responses are far less obvious. She may feel warm and tingly all over, but she won't necessarily associate her arousal with changes in her genitals—or vice versa.[3] As a result, her turn-ons are more diffuse, less defined, less narrowed. However, these differences are slowly changing. An increasing number of women are deliberately using masturbation and fantasy to cultivate more defined, focused erotic preferences.[4]

THE DARK SIDE OF LUST

Turn your attention toward any significant arena of human life—work, creativity, economics, politics, friendship, or family life—and you'll soon come face to face with manifestations of the least appealing and most destructive aspects of human nature. The same, of course, is true of lust. Considering the devaluation and mistrust of lustful feelings in our society, it's not surprising that many people see lust solely as a negative force.

Others try to shield themselves from the dark side of lust by insisting that predatory sexual behavior isn't really sexual at all. It's about power, they insist, not sex. This is the currently fashionable explanation for acts such as rape and incest. But it *is* sex and power and hatred—all rolled up together with the delusion among many predators that their victims actually enjoy it.

The great analyst Carl Jung believed that to achieve psychological wholeness and maturity, each person—and society as a whole—must come to terms with what he called "the shadow," the least acceptable, often denied parts of ourselves. I believe that the same is required for erotic health. It is a terrible mistake to deny or downplay the dog-eat-dog aspects that exist in all human interactions, including sexual ones. Only when we see both the positive and the negative expressions of our lusty impulses can we truly appreciate—with our eyes wide open—all the ways lust can add richness and zest to life.

Lust connects us with our animal passions and brings us closer to primitive energies and motivations, which is precisely why it is so often feared. So it's crucial to realize that lusty urges are most affirming when they are woven into the fabric of everyday life. Conversely, lust is most likely to turn destructive when it is split off from the rest of life, banished to a dark corner where it festers and grows hostile. Lust, by its very nature, objectifies, at least to a degree, but if you experience lust as an integral part of your total self, lusty objectification is balanced by your capacities to empathize with and respect others. And so, for example, while you may fantasize about taking someone sexually against his or her will (or about being taken), you will be able to draw a clear line between fantasy and behavior.

Built into many a lusty fantasy or encounter is a hidden hope for more. In someone who experiences the full range of human needs, fears, and dreams, lust is sometimes the most tangible expression of a desire to reach out, to overcome physical separation and loneliness. How often has a momentary, casual turn-on ignited a desire for another moment with that particular person? More often than you might think.

ROMANTIC ATTRACTIONS

Romantic attractions share with lusty ones a compelling response to a fascinating other. But whereas lust's primary objectives are arousal and orgasm, romantic attractions always include a craving for a mutual passionate bond with the other person. The romantic urge usually aspires to an even deeper goal—no less than personal transformation through the temporary joining of two separate beings. Whereas lusty energy flows from the groin, romantic attractions are experienced as emanating from the heart, although it is usually a meeting of eyes that first alerts you to the possibility of romance.

Historically, psychologists have been reluctant to study romantic love, preferring to leave the subject to poets, philosophers, and artists, who presumably are more at home with the fundamental irrationality of love. In Freudian psychology the search for lost wholeness is associated with an impossible attempt to regain the symbiotic relationship we enjoyed as infants at the breast or even in the womb. To this day many psychoanalysts view romantic desires as regressive, neurotic, and immature. But because the chief concerns of psychoanalysis are the unconscious and the erotic impulses, insightful practitioners have returned repeatedly to the mysteries of love.

Theodore Reik proposed that the romantic urge is motivated by a search for the idealized self. "All love is founded on a dissatisfaction with oneself . . ." he said. "Tell me whom you love and I will tell you who you are and, more especially, who you want to be."[5] According to his view, the fascination that draws us magnetically toward that special other is really just a fascination with our own unattainable perfection. Other students of love insist that the essence of love is the need to overcome the ultimate loneliness of existence, a need implicit in the Platonic myth of the divided self.

I agree with Ethel Person, a contemporary psychoanalyst who breaks with tradition by asserting that the search for love is not merely an attempt to restore a lost connection with the mother. She acknowledges that the unmet needs and unresolved conflicts of childhood all find expression in the romantic adventure, but insists that the ultimate goal is enlargement of the self. The search for love is a creative project, a great act of imagination that is shaped at least as much by where we are going as by where we have been. One thing is certain: the romantic impulse springs from the deepest levels of the human psyche.

THE WAYS OF LIMERENCE

Psychologist Dorothy Tennov insists that our language lacks a word for the unique set of emotions, behaviors, and modes of thought that propel us into love. She has a point. "Infatuation" implies a passing fancy, no big deal. Yet falling in love is life-altering, most decidedly a very big deal. "In love" can refer to any or all phases of romance, from the overwhelming preoccupation of new love to the quiet attachment of an established relationship. Dr. Tennov has proposed that we call the initial, most intense experience of romantic love "limerence" and the attractions that draw us toward it "limerent attractions."

Here's how Dr. Tennov describes the key components of limerence:

- Intrusive thinking about the limerent object (LO)
- Acute longing for reciprocation
- Dependency of mood on the LO's actions
- Inability to react limerently to more than one person at a time
- Fear of rejection
- Intensification through adversity (at least, up to a point)
- An aching of the heart when uncertainty is strong
- Buoyancy (walking on air) when reciprocation seems evident
- Intensity of feeling that leaves other concerns in the background
- Remarkable ability to emphasize what is admirable in the LO and avoid dwelling on the negative[6]

Dr. Tennov reaches the same conclusion as virtually everyone who has made a serious attempt to probe the motivations of love: what we want most desperately is for the beloved to feel as strongly about us as we do about him or her—which seems the only alternative to rejection and loneliness. In Michael's story it's easy to see how everything depends on this reciprocation:

> When I met Lelani it was love at first sight. Not only was she the most beautiful creature I had ever seen, but somehow I knew she was equally beautiful on the *inside*. We had only talked for an hour or so at a party and I couldn't get her off my mind. Even though she was friendly I didn't know how she felt about me or if she would even remember my name. My friend said I was a wreck and he was right.
>
> I called up the woman who gave the party to see if she could help me out. The next day she called to tell me that Lelani thought I was very sweet and handsome. I was dancing around the house like a teenager. Never have I felt so happy.
>
> On our first date we talked intimately over dinner. I wanted to know everything about her and hoped she felt the same. Afterward we spent about three hours kissing passionately in my car—just kissing! It was incredible.
>
> The next time I phoned she didn't return my call and I was heartbroken. I was moping around as if I'd lost my best friend. A few days later she explained that she had to go out of town, that she had wanted to call me constantly, but didn't want to appear too eager. That night we made love with a depth of feeling I hadn't realized was possible. She told me she longed to be with me, *the* words I wanted to hear. That was twelve years ago and we've never been apart since.

Obviously in the throes of limerence, Michael's well-being became entirely dependent on Lelani's responses. A dizzying combination of aliveness and helplessness, the limerent high makes everything pale in comparison to the beloved.

We are likely to experience the onset of romantic love in passive terms, as something that happens *to* us. And so we use expressions like "falling," or being "swept away" by an external force far greater than our own will. This feeling can blind us to a more subtle yet equally important truth: although we cannot control love, we can and do choose it. As philosopher Robert Solomon points out, love isn't just a fall, it is also a leap.[7] Love chooses us, but if we do not accept it and ultimately commit to it, the whole thing fizzles.

LOVE'S OBJECT

With few exceptions, romantic desire is for only one particular person; none other will do. You're fascinated by who he is, what she thinks and feels—especially about you. Whereas lust objectifies, limerence idealizes. When you idealize someone, in a sense you're engaging in an ultra-personal act of appreciation. Ironically, however, idealization can simply be another form of objectification. If someone has fallen in love with you (especially a stranger or casual acquaintance) and you didn't reciprocate, you probably felt very much like an object—a blank screen upon which the person was projecting a private fantasy. And you probably wondered if the real you mattered at all.

Limerence produces a special type of perceptual distortion in which your attention is focused on all that is attractive and desirable about the LO and blots out negative traits or transforms them into endearing signs of the beloved's uniqueness. Those under the spell of early romance are notoriously unable to see the faults of the beloved, even if they are glaring to everyone else—particularly concerned friends and family members.

Love's distortions work two ways. It's not unusual for a lover to feel betrayed and angry when the truth about the beloved eventually overpowers the fantasy. If your original image is too far from reality, you will become increasingly disillusioned and your attraction will wither and die. On the other hand, love's distortions help create a framework for the genuine appreciation that marks the most fulfilling long-term romantic bonds. Even when limerence cools, as it inevitably will, the lover maintains a positive bias toward the beloved, a natural inclination to see what is best.

LIMERENCE AMONG MEN AND WOMEN

The multiple needs that motivate the romantic enterprise—to soothe an aching loneliness, to discover a sense of wholeness in a fragmented world, to compensate for one's doubts and deficiencies by merging with an idealized other, to open up to joy and exhilaration—are all fundamentally human needs that transcend gender. Even so, the ways that men and women approach and experience romance are often quite different. Ignoring these differences can cause painful mistakes.

For the most part, girls grow up steeped in the feelings, language, and imagery of interpersonal connection, closeness, and intimacy. If an infant girl forms a reasonably positive bond with her mother, she will learn to value nurturance, attachment, and the primacy of emotions—lessons she will draw on for the rest of her life. In addition, virtually every girl participates to one degree or another in a collective mythology that includes stories and dreams of love and romance.

By the time she becomes a young woman, responding to social conditioning, to archetypal images of femininity passed down to her verbally and nonverbally, and perhaps also to innate propensities, she will define her identity largely through her intimate connections. Even when she works hard for economic and professional success, she will continue to place a high value on her relationships and expect them to provide a substantial portion of her life's meaning.

It is quite different for most men. A typical boy will also experience his first intimate bond with mother. But whereas a girl readily sustains this bond and builds on it as she experiments with other nurturing alliances, boys must eventually separate themselves from their mothers to establish a masculine identity. This need to "be one's own man," to "go it alone," runs deep in the male psyche. Consequently, for most adolescent males, identity is not established through intimate connections but through achievements, exploits, and tests.

For the male, romantic intensity, if experienced too soon and too strongly, can threaten his budding sense of autonomy. At the same time, his interest in sexuality skyrockets, and with it his fascination with girls, or with other boys if that's where his attractions draw him. But these attractions will soon be contaminated by his need to act cool and unperturbed, no matter how rattled he may feel inside. A man's sense of security remains tentative, fragile, and elusive, virtually impossible to establish once and for all. This is a key factor behind many of the behaviors commonly attributed to male ego.

When a woman falls in love she is likely to be prepared, at least to some degree, for the strong emotions and vulnerabilities that follow. The first limerent experience for a man is typically the same roller-coaster ride of euphoria and risk that it is for a woman, but he enters this new world of emotional intensity with practically no preparation. While it is usually more difficult for a man to leap into limerence, once he does he becomes—and feels—exceedingly vulnerable. Much to the dismay of his partner, he may need to take refuge by avoiding or withdrawing from his LO.

THE DARK SIDE OF LOVE

Love, like lust, has a dark side, a fact our erotic well-being requires that we acknowledge. Love is most likely to show its destructive potential when the romantic impulse is not reciprocated. Both men and women have been known to pursue, stalk, or harass those who will not return their "love." Even when limerent attractions *are* reciprocated, the deep vulnerabilities and cravings they engender cause the worst as well as the best in all of us to emerge when the power dynamics that are integral to the limerent exchange become exaggerated and distorted.

Nowhere is this synthesis of love and power clearer than in the "urge to merge" that appears to be universal in limerence. The challenges and joys of merging are intricately interwoven with surrender and control. On the one hand, without surrender there is no merging. To bask in the transformative powers of the idealized other, you must relinquish control. Love's surrender may or may not be expressed through sexual submission, but it often is—in both men and women.

Inevitably, you begin to feel the need to control your lover. You have crucial emotional interests to protect because your well-being now seems to depend on the actions and feelings of someone else. In addition, you naturally want to express the enlarged sense of self he or she has stimulated. You want your beloved to recognize and respond to your needs, preferably knowing what they are as if by magic. You may resort to manipulation and ultimatums to get your way, strategies not noted for winning affection.

The intensity and insecurity of the romantic situation can push both surrender and control to their destructive extremes. Dr. Person sums up the danger: "When the pursuit of merger is unchecked, the lover becomes either a slave or a tyrant."[8] In her classic book *The Second Sex* Simone de Beauvoir speculates that extremes of romantic surrender may be a natural outgrowth of feminine psychology, especially the inferior social standing of women. At the same time, men—having learned to react to danger with aggression and force—are probably more prone to slip into dominance.

Love's darkest aspects frequently loom largest when limerence is about to die. Occasionally, romantic intensity merely sputters out. More commonly, dying love turns rancid for a time. The same selective and exaggerated perceptions that idealized the beloved in early limerence now reverse and turn him or her into a monster. As love turns to hate, mental energies that once concentrated on adoration are now directed toward fantasies of revenge, sometimes leading to destructive, occasionally murderous acts. Dr. Person eloquently gets to the heart of the matter:

> The deep irrational force in love, so necessary to the projects of transcendence and transformation, may sometimes run amok. This is, of course, why love is so often likened to madness. Passionate love, like all the experiences that open up the self, verges on the borders of self-harm and aggression. In fact, such hazards are intrinsic to all the great creative projects. Rare though the descent from passion to madness may be, its very possibility is the main inspiration for the cautionary approach to love, and for the futile attempt to rationalize and tame it, to declare "rational" mature love as the happy alternative to passionate love.[9]

She believes, as I do, that the romantic experience can only move and change us—for better or worse—to the extent that the usual barriers between conscious and unconscious, ideal self and dark self, are temporarily let down. No wonder the romantic adventure is fraught with risk.

INTERACTIONS OF LIMERENCE AND LUST

Although it's obvious that sex can and does exist without love and that lust often cares nothing about romance, it's not so easy to assert that romance can exist without sex. Although it is relatively rare, some limerent attachments are never consummated in overt sexual behavior. In some cases the emotional risks of romance are so enormous that fear disrupts the physiological mechanisms of arousal or orgasm. Some emotionally intense relationships—most common among women but occasionally seen in male friendships too—have all of the hallmarks of limerence except no apparent sexual desire. Aside from such situations, however, the limerent object is almost always seen as at least a *potential* sexual partner.

But because limerent attractions seek much more than physical gratification, they often pull the desirer in unexpected directions. I've known of several instances in which a person fell in love with someone of the "wrong" sex—that is, not the gender the person normally found erotically appealing.[10] Most commonly, however, the lure of the limerent attraction involves a steamy matching of lusty desire with a profound need for emotional attachment.

ENTER THE OBSTACLES

Strong attractions, whether lusty or romantic or both, are difficult or impossible to ignore. By its nature, attraction is a primary sexual activator. But the erotic equation tells us that we will have a far stronger response to our attractions if they are made more difficult, challenging, or uncertain by having one or more obstacles to overcome. Some arousal-enhancing obstacles work their magic by blocking access to the object of desire—either before or during a sexual interaction.

Notice how multiple obstacles between Grace, a middle-aged divorcee, and her younger lover accentuate their desire and launch them into an incredibly sensuous exchange:

> I had just come out of a very bad marriage and had sworn off guys. Then I met a wonderful man six years my junior. He was very attractive but I was afraid to have sex with him. He seemed so young and I so jaded. And besides, neither of us had enough privacy.
>
> One night after we had gone out to dinner, we came back to my apartment (which I had to myself for a change!) and we did some heavy petting. I began to let my guard down. He was so sweet; he kept saying that he loved me and wanted to show me how much if I'd let him. He said that he didn't have to go home and we could spend the whole weekend together.
>
> Very slowly he began to take my clothes off piece by piece. He picked me up and carried me to the bed. He kissed *every* part of my body. At first I was startled and tried to get up, but he told me to relax and enjoy. So I lay back and let him take me to heights I had never known before. Needless to say, we spent the entire weekend in bed. It's something I still think about. It was also the night I learned all about oral sex.

Here is a clear example of obstacles intervening *on the way to* sex. Grace's low expectations about men, her reluctance to risk again, the difficulty they have finding privacy, and her concerns about their age difference all work to boost their mutual attraction. The slow ritual of undressing symbolizes the peeling away of each obstacle. In fact, overcoming obstacles is so central to the dynamism of this encounter that it's hard to imagine what, if anything, might have occurred if getting together had been easy.

In other situations obstacles are most strongly felt *during* an encounter when differences or conflicts between the lovers generate sparks. Will tells of high drama and wild sex between him and his mismatched girlfriend:

I once dated a girl named Kit who was an awesome fox but unbelievably stubborn and demanding (like me, only worse). Everything always had to be her way so we argued constantly about the stupidest things. I'll never forget one weekend we rented a cabin in the woods. We fought all the way up. I guess I went too far when I called her a fucking bitch because she cussed me out like no one's ever done before.

The strange part was that the sexual tension between us was beyond extreme. The second we stepped into the cabin we ripped each other's clothes off (I mean literally with buttons flying and fabric tearing). It was almost too intense. At one point, I had her pinned down on the floor fucking her brains out. She dug her nails into my back, while we both thrashed around like animals in heat. She said things like, "Fuck me you goddamn bastard!" The battle was depraved, out of control.

Things got even wilder when she started having orgasms, one after the other. She bit my shoulder and slapped me (which really stung). But I continued plowing her until I exploded. We repeated this scene several times during the weekend.

We didn't date much longer. We knew we couldn't handle each other. But right up until we called it quits the sex was always like this, except normally a little calmer. No other woman has ever affected me this way. I have saner sex with my current girlfriend. We get along so much better—so good it's sometimes boring.

Those who, like Grace, must overcome obstacles on the way to an encounter usually engage in tender yet passionate lovemaking once they finally do get together. On the other hand, those like Will, who savor the energy of high-tension relationships, usually engage in rougher, more boisterous sex.

KEEPING A DISTANCE

Because the erotic impulse seeks to bridge the space that separates self from other, among the most effective of all enhancing obstacles is distance—physical, emotional, or geographic. When you're erotically drawn to someone new, the mystery of the unknown creates a realization of distance. This is one reason that visual stimulation is often so crucial to the initiation of sexual interest. Your eyes allow you to reach across the chasm of psychic and physical space, to catch a glimpse of someone who activates your fascination. During flirtation one or both participants play with distance by meeting the other's gaze and then turning away.

Perhaps, like me, you've noticed that flirting takes on a special intensity when circumstances make fulfillment impossible, as when the flirters are about to board different planes at an airport, are rushing off to business meetings, or are with other people and not in a position to respond. Over the years I've heard many married people say, often with consternation, that they were never so attractive to so many people when single. Partly, of course, their lack of neediness places them in a strong, relaxed position, especially compared to those who are desperately searching for someone. However, there's no denying that the unavailability of those who are spoken for boosts their erotic appeal enormously.

The role of distance in keeping erotic interest high is especially obvious when lovers must overcome the obstacle of geography. Those who are forced to endure an ongoing separation wait eagerly for the next chance to see each other. During periods of reunification, erotic sparks are likely to fly. When such relationships survive—obviously the strains are enormous—they typically retain a high erotic intensity long after their living-together counterparts have settled into more comfortable but sexually cooler routines.

In both love and lust, the challenge is to find an *optimal distance*—neither too close nor too far. If you think of sexual arousal as being like an electric spark, you can easily visualize how the size of the gap separating the two poles is crucial. If the gap grows too large, even the high voltage of strong attraction will be unable to jump it. But if the gap becomes too narrow, especially if the poles continually touch, the circuit is completed, making sparks impossible. No wonder successful long-term lovers must find creative ways to balance their closeness with the separateness necessary for erotic enthusiasm.

DRESSING AND UNDRESSING

Ask someone to tell you an erotic story and there's a very good chance you'll hear something about clothing. Not only do our styles of dress portray the image we want to present to the world, but dress is also a concealment, a cloaking of the treasures that lie underneath. In this sense clothing is an obstacle. In the transition from clothed to naked we expose ourselves to each other in more ways than one. The partners uncover each other with a ravenous need to see more, to feel more, and to be seen in return.

Passions sometimes demand that clothes be unceremoniously ripped out of the way. For many people, however, the transition from fully clothed to completely naked is much more exciting when it is slow and deliberate. These men and women often feel equally or even more excited when they're making out partially clothed on the sofa than when they're completely nude in bed. Unfortunately, when a person who savors leisurely undressing is paired with someone who wants to get naked as quickly as possible, frustration is sure to follow.

UNCERTAINTY AND HOPE

Both lusty and limerent attractions can be unilateral; it is possible to be drawn to someone who is indifferent toward you, oblivious to your feelings, or even unaware that you exist. If you're like most people, though, you want at least some response from the object of your desire. Limerence, of course, aches for *total* reciprocation. But Dr. Tennov made a fascinating observation based on her research about the special role of uncertainty and doubt in the formation of romantic attachments. The craving for reciprocation from the limerent object is the essence of the romantic desire, yet she has found that without the experience of doubt, without the risk of rejection, the limerent response may not fully crystallize.[11] In Tennov's words:

> Indeed, too early a declaration on the limerent's part or, on the other hand, too early evidence of reciprocation on the LO's part may prevent the development of the full limerent reaction. Something must happen to break a totally positive interaction. Not that totally positive reactions are without highly redeeming features in themselves; it is only that they stop the progression to full or maximum limerence.[12]

And so it appears that the possibility of unrequited love is a necessary piece of the romantic puzzle. Even in the most idyllic romantic relationships, one or both partners will have moments of doubt about whether they have climbed out too far on a limb, perhaps only to be left there, defenseless and alone. Therefore, those in the throes of romantic passion are inevitably preoccupied with the concepts of "forever" and "always." If you have any doubts about this, notice how frequently these words occur in popular romantic songs.

On the other side of uncertainty lies hope, an essential ingredient in the chemistry of limerence. Without it doubt turns to despair, and despair usually chokes off limerence. And so again we see the interplay, the dance, the dynamism of the erotic equation. Tennov says that "reason to hope combined with reason to doubt keeps passion at a fever pitch."[13]

EMBRACING THE PARADOXICAL PERSPECTIVE

The erotic equation is the foundation for the paradoxical view of erotic life. Obviously, it is not a recipe for terrific sex. It does not predict whom you will find attractive. Nor does it tell you when, how, or which obstacles will heighten your erotic interest. It does remind you of something you've no doubt realized all along: sexual fulfillment is not a straight line from desire to gratification, but is inevitably shaped by struggles, differences, and hazards.

I'm sure you've already been contemplating the interplay of attraction and obstacles in your own peak turn-ons. If you've been jotting down your most memorable encounters and fantasies, now is the time to begin looking at them from the perspective of the erotic equation. First, take note of whatever attracted you to each real-life or fantasy partner. Consider obvious factors such as body type or facial features but also notice which personality traits added to your desire. In what way was he or she similar to you—or different? If you had little or no actual knowledge of the person, what did you imagine he or she was like? What is it about that fantasy that turned you on? If you answer these questions about several different people who have triggered erotic responses in the past, perhaps you will notice recurring patterns and preferences.

Now consider what obstacles added an extra measure of intensity to your peak turn-ons. Were there subtle or obvious barriers between you and the object of your desire? Do you seem to prefer one type of obstacle more than others?

These questions do not demand immediate or definitive answers. They are simply guides to sharpen your observational skills, to help illuminate dimensions of the erotic that might otherwise elude you. Taking time to contemplate how the erotic equation has worked in your life will be particularly valuable as we turn our attention to especially potent obstacles that are fundamental aspects of the human drama. You undoubtedly first encountered them as a small child. And some of them have left an indelible mark on your adult eroticism.

3

FOUR CORNERSTONES OF EROTICISM

The universal challenges of early life provide the building blocks for adult arousal.

On a warm summer afternoon in 1980 I had a peak experience of my own. But unlike The Group's stories of memorable sex, my experience wasn't erotic at all. It was a moment of extraordinary clarity that helped define the course of my work for the next decade. I was soaking up the sun on a beautiful beach, feeling unusually content and carefree. The rhythm of the waves had put me in a trance. It may sound strange, but I was contemplating the elegant simplicity of the erotic equation:

ATTRACTION + OBSTACLES = EXCITEMENT

Only a sexologist would think about such things on the beach. I was appreciating how useful the equation had already been to my clients and me in helping us make sense out of the mysteries of the erotic. Yet I was eager to know more, especially about the obstacles.

Under the spell of the sun questions danced through my mind as if to tease me. Where do excitement-boosting obstacles come from? When barriers help turn us on, do they occur randomly, or are recurring themes and patterns involved? When my clients tell me about their peak sexual experiences, why do their stories seem so similar even though the specific details vary tremendously?

As I pondered I was intrigued by the ever-changing shapes of the swirling clouds overhead. Each new shape flowed out of and then returned to four smaller formations—not just once, but repeatedly. I was spellbound by the apparent repetition even in something as amorphous as cloud formations. I had lost all sense of time, but suddenly the realization hit me that I *knew* the answers to my questions. It was as if the crashing waves, penetrating sun, and swirling clouds had given me access to an inner knowledge I already possessed but had not yet been able to recognize. My understanding did not take the form of words, but a series of provocative ideas began to crystallize in my mind.

The obstacles that intensify arousal are randomly occurring events that cannot be predicted. But because *eroticism is the interaction of arousal and the challenges of living and loving,* the adventure of growing up gives each individual's sexuality a unique shape and texture. Inevitably, each person learns to associate particular kinds of obstacles with heightened excitation. Associations that are sufficiently compelling are likely to be repeated, solidifying the connection still further. Because no two people have exactly the same life experiences, variation is the hallmark of human sexuality—just like the swirling cloud formations.

Although each individual is unique, we all participate in the same realities of human existence. Underlying the events that make each life one-of-a-kind are fundamental life experiences, shared by us all, that involve overcoming obstacles. These experiences—the universal ones—are most likely to find a place in our developing eroticism. This is why the four repeating cloud formations fascinated me so much. They represented the paradox of randomness and structure, of pattern within variation.

Still in a trance I sat up, fumbled in my backpack for a pen and paper, and wrote the words "Cornerstones of Eroticism" without even thinking about it. I then scribbled this list:

Longing and anticipation

Violating prohibitions

Searching for power

Overcoming ambivalence

I returned home feeling slightly dazed. But as weeks and months passed, the four cornerstones stayed with me, leading me to a deeper appreciation of the mysterious ways of the erotic mind. A few years later when I developed the Sexual Excitement Survey, one of my top goals was to look for explicit, clear references to the four cornerstones in the descriptions of peak sexual experiences. If the cornerstones are as central to the human experience as I believed, they should be particularly obvious during moments of peak excitement. As it turns out, unmistakable signs of at least one of the four cornerstones appears in more than three-quarters of The Group's memorable encounters and fantasies.

Because they are woven into the fabric of human existence, I consider the four cornerstones the *existential* sources of arousal-enhancing obstacles. No two lives follow exactly the same course, yet everyone has intimate knowledge of these four essential challenges. And because each cornerstone brings with it obstacles to be overcome, they are ripe for inclusion in our erotic patterns. I am not saying that the four cornerstones are required for enjoyable sex. But they add zest so effectively to memorable encounters and fantasies that without them, eroticism as we know it could not exist.

I believe it is virtually impossible to appreciate your peak sexual experiences fully unless you understand the dynamics of the four cornerstones.[1] Peak turn-ons provide unparalleled opportunities for you to observe the cornerstones in action. During peak moments all the key components of arousal are highlighted, making it easier to see how one or more of the cornerstones actively contributes to the memorability of a turn-on. Once you know what to look for, you can readily see them at work, usually more subtly, in everyday sexual experiences.

As we discuss each cornerstone in detail, begin by noticing to what extent each one plays a part in your peak turn-ons. If you notice a cornerstone recurring in many of your peaks, it probably holds a special place in your eroticism. If so, there's a good chance that you can uncover signs of it in your earliest sexual memories. To help you find out, contemplate two key questions:

1. Think back as far as your memory will take you, to the very first time you felt anything that now, in retrospect, seems even a little sexy or arousing. What do you remember about the circumstances surrounding this earliest experience of arousal?
2. How old were you when you first remember having any sexual fantasies or thoughts? What do you recall about them?

As you ponder the distant beginnings of your sexual self—perhaps writing your thoughts in your journal—don't be surprised if your memories are rather vague. Even fuzzy memory fragments can contain important clues about your erotic development. Notice especially if you can identify any of the four cornerstones in your oldest memories, where your erotic patterns began to form.

CORNERSTONE 1: LONGING AND ANTICIPATION

Part of being human is the ability to picture in your mind something or someone you desire but don't have, or isn't there in the way you want or as often as you wish. This capacity develops shortly after birth and stays with you for the rest of your life. As a child you undoubtedly remember yearning for Mom when she went away or counting the moments until Dad came home. Or perhaps you created an imaginary playmate, someone you could always count on, who would never disappoint or hurt you. When you yearn for someone, the reality of his or her absence or unavailability is the obstacle you seek to overcome by remembering or fantasizing.

According to psychoanalytic theory, the earliest instances of eroticized longing universally occur between the ages of three and five as a desire to possess the parent of the opposite sex. I remain skeptical that these Oedipal urges are anywhere near as universally significant as the Freudians believe, although my research neither supports nor refutes the theory. A great many of The

Group's earliest sexual memories do involve vivid experiences of longing, but only a few—all reported by men—are directed toward a parent. Hank writes:

> I know I was very young when I became obsessed with getting my mother to hold me. Now I recognize that thinking about this aroused me at times. The big problem for me was the fact that Mom was a doctor and rarely at home. I was cared for by a series of nannies. Sometimes I even fantasized that I would be taken to her office and be examined as if I were one of her patients. To this day I find myself wanting more attention from the women I date than they are willing or able to give.

Hank's recollection coincides with my observations as a therapist that sexualized longing for a parent is primarily a compensation for lack of availability. After all, *longing always directs its attention toward what's missing or in short supply.* Keep in mind, however, that as adults we're likely to transfer old feelings of longing toward either parent to whomever we desire now, regardless of their gender.

Most members of The Group recall their first identifiable sexual fantasies between the ages of eight and fourteen, well past the Oedipal stage. Typically, they describe a combination of mental imagery and unfamiliar body sensations—such as warmth and tingling "down there"—when they imagined being close to a favorite movie or TV star or a special teacher, or wistfully yearned for the attention of a someone who seemed highly desirable *and* disinterested. Debbie is one of many men and women who recall such early "crushes." She was about nine or ten when she became infatuated with a teenage boy who lived next door:

> I remember trying to get his attention when he was cutting the lawn with his shirt off. He was very sweet but mostly he ignored me. One summer day he and his family were visiting and I was jumping around in my inflatable pool. I arranged for my bathing suit to sort of "fall off," hoping that he would notice. Everyone just laughed—*not* the reaction I was looking for. It must have been a year or two that I fantasized that he would sneak into my room late at night and carry me

with his strong arms to his bed. I felt so funny when I thought of him. It's now clear to me that I was turned on.

Debbie's attempts to gain the attention of the object of her longing are virtually identical to those seen in adult attractions, although most grown-ups use slightly subtler methods (but not necessarily). Also like Debbie, most young longers imagine someone older initiating and guiding them into a new realm of experience that is as fascinating as it is confusing.

LONGING AND FANTASY

Longing has a unique relationship with fantasy. Whether the object of longing is real or imaginary, I believe that longing *is* fantasy, for both children and adults. When you long intensely, you not only form a mental picture of the one you desire, you can actually *feel* what it was (or might be) like to be close to that person. Without the ability to fantasize, longing simply cannot occur.

Longing, like all acts of imagination, is highly selective. It focuses your mind on the most desirable qualities of a person and ignores or downplays the unappealing ones. If you actually have a relationship with the object of your longing, you look forward to opportunities to be with him or her and relish any communications you may have. The briefest of notes, a moment on the phone, a flower, a knowing glance across a room, or even a secondhand message fragment—any of these can stimulate your desire.

Longing also has a natural affiliation with romantic love. It's difficult to imagine the experience of limerence without the preoccupation that fills the hours while the lovers are apart. What is this preoccupation if not fantasy? Romance novels, enormously popular among women, typically use delayed or interrupted fulfillment to heighten the titillation. Lusty sex acts occur against the backdrop of uncertainty, endless trials and tribulations—all of which make it exceedingly difficult for the lovers ever to get together. When the lovers finally embrace, share a passionate kiss, or make love, their joy is usually short-lived. Soon new obstacles intervene so that yearning can continue.

The most extreme type of longing—falling head over heels in

love with a person who seems to feel little or nothing in return—has been called "unilateral limerence" by sexologist John Money. As a rule such one-sided longing doesn't endure, because the lover who receives no response eventually gives up and lets go, though usually not without considerable pain. Some folks, however, keep hoping for a very long time.

Many people deliberately populate their favorite fantasies with characters they can never have. The imagery is of fulfillment, but the arousal springs from longing, as in Rachel's simple fantasy:

> I have a thing for a guy who plays for the San Francisco Giants. I go to a lot of the games because just seeing him on the field or even hearing the announcer say his name makes me wet.
>
> When masturbating I pretend he's my boyfriend and completely adores me. It's so simple, really. We just make love and I feel very close to him.

Whereas Rachel's fantasy is the product of pure imagination, other experiences of longing are inspired by actual events that are unlikely to happen again. Many of The Group's most compelling fantasies relive, typically with embellishments, wonderful past encounters with former lovers or steamy, once-only trysts with strangers. Some longing fantasies acquire their power from missed opportunities or might-have-been or almost-were experiences. Sammy, a gay man now approaching forty, continues to relish an encounter that never quite happened when he was just fifteen:

> I invited two buddies to stay overnight. I was *extremely* attracted to one of them who sat next to me in class, where for months we had touched knees. First we took a swim in our pool. When we changed clothes, I saw him nude. He was dark, physically mature, with a beautiful dick.
>
> Finally he was in my bed but, unfortunately, we weren't alone. When he got up to turn off the light he had a big hard-on. I wanted it so much I was going out of my mind. We started touching a little, but I was worried about the other guy so I held back. Fuck!

> After he fell asleep I leaned over and kissed him on the lips. I was awake all night horny as hell. I *still* want him!

Situations in which we actually get a taste of what we crave, but not total fulfillment, are particularly likely to stay with or even obsess us. Longing reaches its zenith under conditions of partial or intermittent satisfaction. If expressions of interest and attraction are interspersed with signs of detachment—maybe the desired one pulls back, turns cold, or goes away for a while—the result can be a frenzy of desire. Anyone who has ever become involved with someone who already had a primary relationship knows how just an occasional crumb of interest or reciprocation acts as an aphrodisiac. Actual moments together take on special significance.

ANTICIPATION: SHORT-TERM LONGING

Longing and anticipation are variations on the same theme; both draw energy from the gap between desire and the reality of the moment. The difference is that longing must usually overcome formidable barriers. Her lover lives in another town so they see each other only occasionally. His girlfriend is involved with someone else so he waits by the phone for calls that rarely come. With anticipation, the wait is not nearly so prolonged or painful because fulfillment seems relatively near. In a state of yearning you are intensely aware of the experience of being without, whereas anticipation is almost entirely focused on the goal of being together.

Thirty-nine percent of The Group's peak encounters contain references to desiring an absent or unavailable partner, or anticipating the encounter itself or some specific moment within the encounter. Far fewer—18 percent—mention longing or anticipation in their fantasies. For many people fantasy is an opportunity to use their imaginative abilities to guarantee gratification. Yearning enthusiasts, however, often prefer to build up their arousal gradually by visualizing an extended seduction or some other circuitous path to satisfaction.

Overall, longing is significantly more common among women than among men, with lesbians the most likely of all to mention it. One reason the women report greater longing is that they are more likely to have romantic feelings toward their partners. While anticipation can be a component of either limerence or lust, serious longing is definitely associated with romance.

In actual practice it's virtually impossible to make a clear distinction between longing and anticipation. The dance of longing and anticipation is obvious in a story told by Frank, the only member of The Group who mentioned his wedding night as the setting for unforgettable sex. Because of Frank's work, he and his fiancee had to be apart for nearly six months prior to their wedding, forcing them to make most of their plans by phone:

> We decided we would stay overnight in a hotel before our honeymoon. I requested just one thing—that she wear a garter belt. She laughed but admitted that she too had some special things in mind. The consummation of our wedding was on my mind constantly.
>
> During the ceremony I kept looking at her, thinking how beautiful she was, that we would be together at last. Later in our hotel I was so aroused I could hardly keep my pants on. But we undressed each other slowly, very tenderly, taking all the time we needed to fully enjoy every moment. Our sex that night was the best I've ever had. Why? I'd have to say it was the celebration, our deep feelings for one another—and being apart for months didn't hurt either!

Notice how both long-term longing and short-term anticipation work together as erotic intensifiers. Who wouldn't feel strong anticipation under such circumstances? But years before he ever met his fiancee, Frank already had a penchant for anticipation, a fact that is obvious in his second favorite encounter, remembered from more than ten years earlier. A very different kind of encounter, casual rather than romantic, it still builds energy from anticipation:

> I knew a girl who liked to have sex a lot and was totally uninhibited and multiorgasmic. We'd meet at the oddest times and weirdest places. The one that stands out most happened on a warm summer evening when we were sitting on a

golf course smoking a joint. Suddenly she stood up, stripped off her clothes, and took off running down a fairway. She was giggling and shaking her ample breasts at me in the moonlight.

I tried to catch her but she raced ahead, taunting me that I could have her if I could catch her. Well, I bolted off after her like a man possessed. When I finally caught her—which wasn't easy—we tumbled to the grass and made love with overwhelming passion.

Even subtler forms of flirting draw much of their power from the dynamics of anticipation. Flirting only works when there is an awareness of distance—a gap—between flirter and flirtee. Erotic fascination is activated by the possibility that the gap can be bridged. The more prolonged and intense the flirtation, the greater the anticipation and the more powerful the desire.

TEASING AND ANTICIPATION

Both longing and anticipation work their magic in the time and space before sexual contact. Once sex begins, anticipation usually recedes as attention focuses on the pleasures of the moment. Teases are the exceptions because they bring anticipation *into* an encounter, even after all barriers to contact have been overcome. Skillful sensualists learn to touch in ways that build rather than reduce anticipation. Lucky recipients like Beatrice, age fifty-four, sing their praises:

My husband has a tendency to go directly for what he wants, even if it's me. I'm glad that he still wants me after all these years, but I wish he would take his time. For some reason, one morning he decided to do just that. Without a word, he began stroking my back and then my butt and legs. I was in heaven and really getting juicy.

He did things like run his finger up my thighs, gently brushing against my pubic hairs, which raised goose bumps across my body. When he turned me over he touched my nipples so lightly with his tongue that I was practically screaming for more, but he would only do it for a moment and then move on. I have no idea what got into him that day, but he teased me into one of the most intense orgasms of my life.

Experts typically explain the desire of so many women for extended foreplay on the basis of the slower rhythms of female sexual physiology. Beatrice's story reminds us that there is often much more involved. A slow, sensuous buildup of arousal is among the few ways established couples can experience the miracles of anticipation. Unfortunately, men are sometimes reluctant to go slowly because they associate prolonged anticipation with being thwarted or put off—a common and frustrating experience in men's sexual histories.

LONGING AND FULFILLMENT

The erotic significance of longing is impossible to deny. Yet one of the great paradoxes of erotic life is that even though longing craves fulfillment, fulfillment dampens longing. In some instances longing evaporates immediately after the last barrier to access dissolves. An extreme but not unusual example is a pair of coworkers who maintain a strong but "impossible" sexual curiosity for years, only to be permanently satiated by a single chance encounter—perhaps even a very good one. Some erotic fascinations are founded on unavailability and simply can't survive without it.

In more complex attractions, longing normally subsides during and after a passionate encounter, but returns once the lovers part. Yearning renews their passion—at least for a time. However, the predictable togetherness of living-together partnerships often makes longing increasingly difficult to sustain. For more than a few lovers, the demise of longing is a serious impediment to ongoing desire.

Yet even couples with relatively few opportunities for longing can still benefit occasionally from its aphrodisiac effects. Sometimes even a brief separation caused by independent travel or the emotional distance created by an argument can be remarkably effective at rekindling longing. In fact, almost 10 percent of The Group's peak encounters are reunions following such separations or fights.

Other subtle manifestations of longing aren't necessarily dampened by togetherness. I've worked with many people in couple's therapy who describe yearning for certain emotions or behaviors that remain out of reach despite the existence of a committed relationship. One woman repeatedly explained to her husband (he thought she was nagging) how much she craved more emotional closeness with him. On those rare occasions when he disclosed intimate feelings to her, she felt incredibly excited. Her fondest memory was one night when they cried together and then made passionate love. As therapy continued, he revealed how much *he* longed for those equally rare instances when she totally surrendered to sex. He hadn't realized that his own emotional openness was the magic potion that brought her passions to life.

A relatively rare form of longing operates according to a different set of rules. Some people yearn for love so profoundly and for so many years that they never forget that experience, even when they eventually do form a close, intimate bond. Repeated fulfillment doesn't reduce their longing, but rather reminds them of how lucky they are to have beaten the odds and found a loving mate. They have developed a self-generating, longing-based erotic system that enables them to nourish a fascination with their partners for years and decades. They retain the aphrodisiac effects of longing by holding it in memory, where it sweetens their fulfillment.

CORNERSTONE 2: VIOLATING PROHIBITIONS

Every society tries to limit sexual behavior. Not only do these cultural restrictions define and enforce the ideals and mores of the community, but they also have another function that is not consciously intended: they provide ready-made barriers that anyone can use to intensify his or her turn-ons. We're all born with the capacity for arousal, and sooner or later (usually sooner) we'll experience it. But what happens when we sense (or know) that adults don't want us to feel this way?

The erotic equation predicts that those who grow up in sexually restrictive environments are almost certain to discover the erotic potential of breaking the rules. If you can recall any titillating childhood adventures—such as playing "show me" or "doctor," being fascinated by pictures of semiclothed people in catalogues or *National Geographic,* secretly looking up sexy words in the dictionary, or discovering the parts of your body that weren't supposed to be so pleasurable—you probably had two contradictory reactions. At times feeling naughty, dirty, guilty, or afraid of punishment may have restrained you from further experimentation. On other occasions these feelings might just as well have added an extra charge to your activities and made you want to repeat them.

A fusion of arousal and rule-breaking when you're young dramatically increases the odds that you'll retain in your adult eroticism a tendency to be excited by violational behaviors and fantasies. I call the aphrodisiac effects of violating prohibitions the *naughtiness factor.* Not surprisingly, it is especially pronounced in societies that seek to block most, if not all, expressions of childhood sexuality. And such societies dominate the modern world.

On a more personal level, you're certainly not alone if you remember adolescence as a period when the link between nervous excitation and breaking the rules loomed especially large. Tina still thinks of an encounter from her teen years as among her most exciting, even though she's now almost thirty, married, and pregnant with her second child. Her story captures perfectly the aphrodisiac qualities of sexual rebellion:

> My boyfriend took me home after I watched him play in his band. Earlier that evening we were messing around so we were still pretty jazzed. We parked in my parents' driveway. The backseat was filled with his band equipment so he crawled over the stick shift and sat on top of me. He undid my bra and my pants as I reached inside his pants. We kissed and touched until we couldn't stand it anymore.
>
> We switched places so I was on top. My pants were all the way off and his were around his ankles. With my legs sprawled apart, I sat on top of him so his penis could reach deep inside of me. There we were, bobbing up and down, kissing passionately and looking over our shoulders in case

> someone came outside to greet us—it was all very exciting. It was daring to make love with my parents inside. They *hated* my boyfriend because they thought he was trying to seduce me!
>
> It only took a few moments for both of us to come, calm down, and put our clothes back on. We laughed and kissed as he walked me to the door and said good night, both of us looking quite smug.

You can see how the push-pull of inhibition versus titillation can be a high-stakes juggling act. These two are courting disaster *and* having a terrific time. Fewer adults are forced into such classic predicaments, but the thrill of naughtiness is ageless and timeless. Thirty-seven percent of The Group's peak encounters contain similar references to the excitement and risks of violating prohibitions. They often use words such as "raunchy," "sleazy," or "trashy" to highlight the forbidden quality of their behaviors. For grown-ups, two situations are most likely to activate the naughtiness factor: (1) a risk of getting caught or discovered and (2) an attraction to disapproved partners.

BEYOND PRIVACY

A common feature of adolescent sexuality is the quest to find a corner of private space hidden from the watchful eyes of disapproving adults. Once we become adults, most of us choose to have the vast majority of our sexual encounters in private. Nonetheless, the adolescent struggle for autonomy and privacy sets the stage for adults to be aroused by the thrill of sneaking, hiding, or risking discovery. This is why even married partners can get an extra charge from making out in the car. They're not *really* afraid of being caught or punished, but the ambiance of naughtiness increases the erotic tension—not so much, however, that it gets in the way.

Hillary, a fifty-year-old fashion designer, wife, and mother, describes the dilemma posed by a family vacation and the sparks that resulted. Much to their dismay, she and her husband had rented a too-small cottage in which it was impossible for them to get away from the kids. "We were forced into celibacy," she lamented. After four days of frustration she and her husband finally took a walk alone in a park:

> It was a wonderfully romantic night. The park was beautiful by day or night. In no time we were kissing and hugging, all the while looking anxiously around to see if anyone was coming. I was frustrated *and* completely excited. My husband suggested we find a sheltered place to go at it. At first I thought he was kidding, but I soon found out otherwise.
>
> We picked our hiding place, lay down on our jackets, and had a marvelous time playing. One breast sticking out of my half-open blouse really got my husband's attention. And when I opened his zipper and his rock-hard penis bounced out, I almost came on the spot. But we never stopped listening for approaching strangers.

It was innocent fun. Most stories of risked discovery are similar to this one, containing only a hint of real danger. For most people, too great a preoccupation with discovery generates so much anxiety that it ruins everything. There's considerable variation, however, in how individuals evaluate the level of risk and how much of the associated anxiety they can tolerate. Craig, a gay man in his late twenties, is quite bold in his pursuit of forbidden thrills:

> I was waiting to board the train when I spotted a handsome businessman. I felt aroused by his provocative eye contact and hoped he was on my train. He followed me and chose a seat directly across from me. We shared glances while both pretending to read our newspapers. This continued for about ten minutes until he started fondling himself slightly. I was going out of my mind.
>
> Finally I stared at his crotch as I played with mine. I couldn't believe this was happening on a train! He motioned for me to come over and we groped each other's hard-ons through our clothes. We unzipped and masturbated each other until we both came in our underwear. The excitement was overwhelming, but we never spoke a single word.

Similar stories of semipublic sex, sometimes with strangers, sometimes with lovers, are told by both men and women, regardless of their sexual orientation. Gay men are most likely to mention them. Women are more likely to *fantasize* such encounters than actually to have them. For most women, anonymous, real-life lusty sex feels too dangerous to enjoy. When women do tell of semipublic sex with strangers, they virtually always emphasize features of the encounter that promote a sense of security, such as a remarkably gentle, attentive, or nonthreatening partner.

FORBIDDEN FRUIT

Even though it may not fit your ideals about love and sex, the unvarnished truth is that partners who catch our attention by virtue of being inappropriate or forbidden are often among the most magnetically attractive. You can readily see this principle at work in illicit affairs. While you may find the idea threatening or disturbing, you are no doubt aware that the forbidden nature of an affair gives it exciting elements not present in long-standing, committed relationships. The newness and greater opportunities for longing provided by an affair combine with the belief that the affair is wrong to produce a strong erotic charge.

Notice how JoAnn enjoys both sources of naughtiness—a forbidden partner and the risk of discovery—in an affair with another woman, whom she refers to as "M." But her partner isn't just any woman. She's the ex-partner of JoAnn's own current lover:

> I found it very arousing to know them both intimately. I felt "between" them, so to speak. It was nasty to rent a private room in a hot tub place, sort of like paying for sex by the hour. I imagined that everyone could tell instantly why we were there. Believe it or not, I'm quite conservative and traditional. So it was stimulating to feel unlike my normal self. "M" was dominant—which I loved. We had rough and wild sex with lots of screaming, biting, and scratching. I enjoyed making a lot of noise and letting myself go.
>
> She sat on the edge of the hot tub and I went down on her. She is the type of woman who ejaculates when she comes. There I was, kneeling in the tub, giving her head, while she was coming all over my face and into the water. I

felt like a total slut. We made a very sexy mess and it was an added bonus that we didn't even have to clean up.

The imagery of passion gone wild permeates JoAnn's story. She paints a picture of sleaze and dirtiness with big, bold strokes. Following a more subtle approach, Helena, age forty-nine, makes her partner *seem* forbidden by associating him in her mind with someone else:

I was alone in Mexico, a tourist. While visiting an archeological site, I struck up a conversation with an architecture student who also wrote poetry, studied Yoga, and was a runner. His combination of energetic youth and lively intelligence attracted me powerfully. Also, he was close to the age of my younger son (early twenties), which provided a thrill of an almost incestuous kind.

We walked back to town—about five miles—where I smuggled him up to my room. We made love several times during the next few hours. Rarely have I had a lover more enthusiastic. Yet he was tender and thoughtful of what I wanted and how I wanted it.

Tom's encounter also has an element of age inappropriateness. When he was twenty-six he was hired to lead a tour of high school seniors. The rigorous training was full of warnings never to become involved with the students and that doing so would be grounds for immediate dismissal. The limits were quite appropriately strict, but they also helped eroticize the atmosphere:

Many of the students were away from home for the first time, eager to push their freedom to the limit. While other leaders tried to clip the wings of their kids, myself and two other leaders adopted a more liberal philosophy. We gave them condoms and asked only that they use them if they chose to have sex.

One student, Beth, had just graduated from high school, was an aspiring marine biologist, and really sweet. She frequently sought me out to talk into the wee hours of the morning. She developed a strong crush on me but I was reluctant to let things develop because I didn't want to take advantage of my position.

As the summer progressed I became convinced that her feelings toward me went beyond a crush—as did mine. We spoke about this and we agreed we would have to wait until we got home. Each passing day became an exercise in excitement and frustration. Each touch, each comment was laden with sexual overtones.

After almost five weeks of this foreplay, my fellow group leaders, four of whom were having torrid affairs with each other, invited me for a midnight skinny-dip in the sea. They encouraged me to invite Beth. Perhaps it was the moon shining on Beth, the soothing waves, or just the tremendous sexual tension built up for so many weeks, but we kissed and couldn't stop. She climbed on top of me and allowed me to enter her. It was so warm and felt so right. We were afraid of being spotted by the other leaders so we remained quiet. She climaxed twice and I once.

I will never forget this experience. With the exception of one other "slip," we waited until we returned to the States before having sex again. We spent the transatlantic flight explaining, in minute detail, what we planned to do to each other when we got home. These lurid descriptions were sufficient to bring her to climax twice without any touching at all.

Longing and the naughtiness factor join forces here for high excitement. Notice your reaction to this story. Did a slight sense of shock contribute to your involvement? The protagonists have unmistakably crossed a line that most people, including Tom himself, would agree has a legitimate and proper function—to protect the students. But my guess is you're not that upset. For one thing, the rules broken here were not sufficiently grave to scandalize many people. Then too, the situation was highly romantic. And love is widely seen as an acceptable reason for breaking rules. Had the storyteller been motivated solely by lust, we might feel quite differently about him.

PUSHING THE BOUNDARIES IN FANTASY

The naughtiness factor obviously draws its power to excite from the interplay of desire and the awareness of limits. Initially, restrictions are imposed by external powers that be. Gradually, though, each person internalizes at least a few of these boundaries as personal values. Consequently, most of us are comfortable placing limitations on our sexual behavior—even if we don't always honor them—because they seem rational, necessary, or right.

In the realm of the sexual imagination, however, we are more likely to throw off the restrictive effects of rules and taboos, including heartfelt ones, sometimes getting a thrill out of shocking even ourselves. Consequently, many people go much further in their fantasies than they would in real life. After all, no one has ever been harmed by a fantasy. Actually, somewhat fewer of The Group's favorite fantasies contain clear references to violating prohibitions than their real-life encounters—29 percent versus 37 percent. But those who enjoy the naughtiness factor in their fantasies often trample on taboos in ways they never would in reality. You can see how Brian conveniently ignores a host of prohibitions and practical considerations in his most erotic fantasy:

> I'm waiting for my doctor's appointment when I notice how incredibly sexy the receptionist is. There is a mirror on one of the walls and in it I can see between her legs. I start to feel an erection building and I halfheartedly try to cover my crotch with the magazine I'm supposedly reading. She opens her legs wider and sticks two fingers in her pussy and licks them clean with her tongue. She comes over and tosses aside the magazine and proceeds to undress me. We get into a sixty-nine position when a nurse comes into the scene, already naked. I fuck her in the ass as she goes down on the receptionist's pussy.
>
> If I really want to get sleazy, I'll have an innocent pubescent girl unexpectedly run into us on her way out of an examining room. She's so overwhelmed and excited by what she sees that she lingers to watch. Soon she unbuttons her blouse and presses one of her small, firm breasts into my face, begging me to devour it. The fantasy culminates with us making it everywhere in the office. Then, totally wasted, the receptionist hands me a card with my next appointment date.

The fact that fantasy has the potential to release us from all social, moral, and pragmatic constraints is among its most useful features. But exactly *how* the imagination lets us revel in the unacceptable is complicated, if not rather mysterious. On the one hand, we vigorously disregard the taboos we violate during fantasy, pushing them out of our minds, declaring them irrelevant. If, however, we totally lose sight of what we're violating, the fantasy's entire reason for being suddenly crumbles. It seems as though we must forget *and* remember at the same time—quite an elaborate juggling act, to say the least.

It's particularly difficult for those who genuinely believe in the very taboos they wish to flaunt. In Judy's favorite fantasy, she struggles with her attitudes toward prostitution, something she's quite passionate about in two very different ways:

> Ever since I was about fifteen I've fantasized about being a prostitute. I was always supposed to be "good," but prostitutes claim the right to be blatantly sexual. As a hooker, I relish my seductive walk, whorish clothes, and dirty talk. I imagine a man slowing down for a look at me. If I like what I see, I ask if he's in the mood for action. Sometimes I'm a streetwalker and we do it in his car or a fleabag hotel. Other times I'm a sophisticated call girl catering to rich businessmen. But I'm always in control, totally sexual, and I don't give a damn about what anyone thinks.

Judy goes on to explain her evolving feelings toward her hooker fantasies:

> As a kid I felt concerned about my fascination with whores. Maybe I really wanted to *be* one—a horrifying thought. Recently I became involved with others in my community to drive the street hookers out of our neighborhood. I feel very strongly about this issue especially since a couple of kids found used condoms and needles in the park. More than once I went home from one of these meetings and masturbated in the bathtub (my favorite spot). And what did I fantasize about? Prostitutes, of course! I felt like a terrible hypocrite. But then I realized that my thoughts are my own business and totally unrelated to the real world. I still feel a certain uneasiness about my fantasies, but I think I like it.

Those who enjoy the naughtiness factor want and need the very rules and limits they get such a kick out of challenging—one more erotic paradox. Without boundaries to push against, there is no joy in naughtiness. If hookers roaming the streets were as meaningless to Judy as newspaper boys, they could no longer serve her as symbols of wanton lust.

During the sexual revolution of the 1960s and 1970s, when many restrictions from the past were cast aside, it became increasingly difficult to feel naughty. Some of my clients during that period complained about having too much freedom. I'm convinced that the attempts of many individuals and groups to shore up traditional values during the 1980s and 1990s haven't come simply from antisexual, moral, or political motivations. How many, do you suppose, want tighter restrictions on sex because they miss having strong prohibitions to push against? Could a desire for forbidden pleasure be an unconscious source for "antisexual" attitudes? Consider the following evidence.

CATHOLICISM, SEXUAL ORIENTATION, AND THE NAUGHTINESS FACTOR

The thrill of violating prohibitions is clearly evident in all segments of The Group, regardless of gender, sexual orientation, or religious affiliations. But whereas The Group as a whole mentions the naughtiness factor in 37 percent of their peak encounters, three subgroups are especially likely to do so: (1) 69 percent of those raised as Catholics, (2) 64 percent of lesbians, and (3) 41 percent of gay men. I believe the primary reason is their shared experience of growing up in sexually repressive environments.

Those raised Catholic, even those who no longer think of themselves as members of the church, still exhibit a passionate involvement with the naughtiness factor—evidence that adult erotic patterns are launched early in our lives. It's just as the erotic equation predicts: the more consistently disapproving messages surround us when we're growing up, the greater our affinity for the forbidden will be when we're adults.

Most Catholic children are strongly discouraged from pursuing their natural sexual curiosity as kids. Nevertheless, most get the message that sexual expression will become at least acceptable in the eyes of God once they grow up and marry. In contrast, children who will become gay learn that the most fundamental feature of their eroticism—whom they're attracted to—is and always will be totally unacceptable. To make matters worse, homophobic attitudes take root long before people know anything about their own sexual orientation. Once attitudes are internalized in this way, a conscious effort is required to change them. Coming out is so important to gays and lesbians because it begins the necessary shift in self-perception.

Even after gays and lesbians accept themselves, many retain deep feelings that gayness is inherently bad, if not downright unnatural. The only advantage to feeling this way is the extra erotic kick it can provide. Consequently, gay men are particularly likely to think of their role within the dominant culture as that of sexual outlaws. They reject the very rules that exclude them and embrace an ethic that emphasizes sexual freedom. Quite a few straight and bisexual men have similar feelings about themselves. The role of outlaw, sexual or otherwise, is more easily adopted by males. Lesbians, like women in general, may enjoy breaking the rules, but usually do so less confrontationally—with, of course, many exceptions.

SELF-ESTEEM AND THE NAUGHTINESS FACTOR

Common sense suggests that regular sexual rule-breaking carries with it the risk of feeling guilt-ridden and ashamed. Especially if early sexual prohibitions are accompanied by threats of punishment or withdrawal of love, the result can be a deep sense of shame about even garden-variety sexual desires. What usually isn't so obvious is that violating prohibitions can provide avenues for self-assertion and affirmation, which *do* contribute to self-esteem. The importance of establishing one's own right to decide is clear among adolescents, even when decisions are dictated by the shared ideals of the peer group rather than the individual.

Once we become adults we define ourselves less through our rebellions than through our accomplishments, values, and relationships. But most of us retain an urge to demonstrate our superiority over the rules that continue to restrict us. Perhaps this is why encounters and fantasies with a flavor of violation so often leave the violators with a sense of self-validation or even pride. Of course, positive reactions like this are easier if the limits and values being violated are somebody else's rather than one's own.

CORNERSTONE 3: SEARCHING FOR POWER

The story of childhood is, to a large degree, the history of our attempts to move from the abject powerlessness of infancy toward a clear and strong sense of self, a self capable of standing its ground in an indifferent and sometimes hostile world. Without some success in our search for power, we would have to live—if we could survive at all—in a state of perpetual submissive dependency.

Throughout your life two fundamental strategies have been available to help you cope with or overcome powerlessness. The first involves direct action. By the time you were two you had already discovered ways to assert your will, probably including stirring up a fuss to get your way, threatening or using retaliation when you didn't, or staging a sit-down strike when all else failed. As you've grown you've added additional strategies, including, with luck, the ability to express your wishes plainly and assertively.

A more subtle and indirect approach can still provide a semblance of control for the powerless. Have you ever noticed how submitting to a dominant other sometimes allows you to join or even coopt his or her control? Highly refined surrendering can give "powerless" practitioners nearly total control. Negative examples of this are regularly acted out by "helpless victims" who become tyrants, demanding total compliance and devotion. But indirect routes to power are by no means intrinsically negative. Both children and adults regularly rely on indirect strategies when they must deal with people in dominant positions—a parent, a bully, a teacher, a boss, or maybe even a lover.

Your search for power, regardless of which strategies you use, always involves overcoming the obstacles created as opposing wills collide. When actual or fantasized power dynamics intersect with experiences of arousal, as they often do beginning early in our lives, the erotic equation predicts that our responses might well be intensified. Consequently, long before many of us reach adulthood, subtle or dramatic themes of dominance and submission have become established as reliable turn-ons.

Twenty-eight percent of The Group's peak encounters contain obvious references to at least mild dominance or submission or both. The percentages grow dramatically, however, if we consider The Group's favorite fantasies. Over half of the women (56 percent) and somewhat fewer of the men (44 percent) make clear references to power in their fantasies, a considerable increase for both genders. I was particularly surprised that women are twice as likely to focus on power in their fantasies as in their real-life encounters. Lesbians accentuate this trend, with 83 percent mentioning power dynamics in their fantasies.

MAKING THE MOST OF POWERLESSNESS

Now is an appropriate time to review again your childhood sexual feelings and fantasies. What were the power relationships between you and the objects of your earliest attractions? When I ask therapy clients about this they frequently remember responding to vague images of dominance and submission in movies or on TV. Most notable are the scenes in which someone is tied up, taken away, or held captive. Similar images are common among The Group's earliest sexual fantasies. Marla, now middle-aged, explains how a fantasy she first remembers having at age twelve contains elements that still entice her today—longing and surrender in a romantic setting:

> I am kidnapped by a dark, handsome man. He puts me on a boat and sails us to an island in the tropics. He builds a small bamboo hut and lays out a blanket where he undresses me and then himself. He ties my hands together and gives me oral sex and drives me utterly insane. Afterward we hold each other in the cool breeze.

Notice how *he* is the source of all the action. But notice also that *she* is the primary recipient of the attention and pleasure—after all, it's *her* fantasy. It could be argued that a girl's early fantasies of sexual surrender are part of her internal preparation for the submissive role she will later be expected to play. After all, the most familiar images of male-female sexual interaction include at least mild domination by the man along with a complementary yielding by the woman.

If, however, early submission fantasies help prepare us for adult sex-role behavior, many of the men in The Group appear to have been studying the wrong scripts as boys. When men remember images of power in their earliest fantasies, as they often do, they're just as likely as women to be submitting to a highly desirable but more experienced and powerful other, as in Juan's fantasy:

> When I was in fourth or fifth grade, I had a crush on my teacher, Miss Peters. I would fantasize that I did something bad (even though in reality I tried to be her favorite) so I had to stay after school. I imagined she took off my clothes to punish me, but I didn't mind a bit. I wanted her to touch me. I especially liked the idea of being forced to sit under her desk while she graded papers, waiting for her to spread her legs so I could sneak a peek at her panties.

With few exceptions, when The Group's earliest fantasies involve power roles, the fantasizer, whether a boy or a girl, is being guided, coaxed, or forced into sensual or sexual experimentation. Yet such fantasies are virtually always described as pleasurable, with the frequent exception of guilt afterward. In my view, we first discover the erotic potentials of receptivity and aggression in the powerlessness, especially concerning sexual matters, of our youth.

PARADOXES OF POWER

Power positions in sex are often described as "top" (the forceful, aggressive initiator) or "bottom" (the receptive, yielding responder). At first glance it may appear obvious who's playing which role. It's commonly assumed, for instance, that the inserter in intercourse—vaginal or anal—is the top, while the insertee is the bottom. Likewise, a person being "done" or pleasured is seen as bottom because the "doer" is more active. Perceptions shift, however, in male-male encounters; the receiver of oral stimulation is usually seen as the top because he is assumed to be in a more manly position.

When people describe the subjective experience of a top-bottom encounter, there's hardly anything obvious about who's in control. I have consistently observed that whenever people engage in sexual power exchanges voluntarily and enthusiastically, whether they play the role of top or bottom, they feel an enormous sense of powerfulness and validation. Peter, a construction worker in his mid-thirties, demonstrates the paradox of empowerment through submission, as his beautiful and aggressive girlfriend teaches him a thing or two:

> I had just stepped out of the shower when she rang the bell. I wrapped a towel around my waist, invited her in, and followed her to the couch. I felt excited and vulnerable to be nearly naked while she was fully clothed. Tension was rising.
>
> She said, "If you're not careful I'm going to rip off that towel." I liked her taking control, but I played it cool (I knew I was driving her wild). Soon she *did* rip off my towel. I was totally naked and she was still fully dressed. There was something completely unnatural about this but also very satisfying.
>
> She took *total* control and did something completely out of character. She turned me on my belly, draped me over the couch, stuck her finger up my asshole, and masturbated me with her hand. It was as if my whole body became a giant penis and she was massaging the whole thing, inside and out. After orgasm, I shivered for ten or fifteen minutes while we held each other. I can't remember *ever* feeling more alive than I did that night.

Peter's memorable encounter reveals a fundamental truth about power and eroticism when positively intertwined: *a forceful partner demonstrates with his or her passion the value and desirability of the one who submits.* The reverse is equally true: *a submissive partner demonstrates through his or her surrender the irresistible erotic powers of the aggressor.* Both top and bottom feel strong and affirmed. Ultimately, control resides within neither alone, because the energy is generated by their interaction.

POWER AND RESPONSIBILITY

One of the advantages of at least appearing to surrender to a more powerful other is that we can disclaim responsibility for what follows. I'm intrigued by how frequently sexual submission is used, especially by females, to circumvent the incredibly intense messages warning them not to be sexual. Marcie, a marketing executive in her late thirties, expresses her insights as she both describes and analyzes her own favorite fantasy:

> I am a plaything at an exclusive men's club. Any time, anywhere, I *must* do what I'm told. Sometimes I must masturbate in front of a whole room full of men. Sometimes I must not react while every part of my body is being stroked and caressed. Then there are the contests in which I see how many men I can make come in an hour—or how many times *I* can orgasm.

Marcie goes on to explain how much she enjoys the fact that she has no control, contrasting this with her actual life filled with the responsibilities of motherhood and a high-stress job. She allows us into her thoughts:

> I always find it ironic that in real life I never willingly let men have their way with me or put me down. Yet my fantasies all revolve around my lack of control. I think it comes from my old belief that "good girls don't." If I have no say over what is happening then I'm not to blame for my enjoyment. I'm just following orders.

Max, a bisexual man who has a primary relationship with a woman as well as occasional trysts with men, explores similar themes of responsibility from a man's perspective:

> Sally and I have a terrific sexual relationship filled with passion and experimentation. But there are definite limits. Maybe it's because I'm a guy, or maybe because I'm big and muscular, but Sally expects me to take charge, especially when we fuck. I've told her I like to be passive sometimes and she sort of tries to take control, but I can tell it's just an act for her. I like feeling powerful when she surrenders to

me, but I *don't* like controlling everything all the time.

I've always had a special kind sexual interest in men. I recently met a bisexual guy who's almost as large as me and who gets off on being on top. I see him once in a while, but one time was the absolute best. I had confided in him that I fantasize being tied up. Some time later he used rubber strips to tie me to the bed frame.

The rush of excitement was incredible as he stimulated me relentlessly. He sucked me almost to the point of coming and then backed off. I was surprised how I enjoyed a little pain when he pinched my nipples, tugged on my balls, or rolled me partway over and slapped my ass. The high point for me was when he got out a rubber and lube and—after loosening me up with his fingers for a good half hour—fucked me masterfully while he jacked me off. It was *better* than my fantasy.

Max expresses a complaint similar to those of several straight male members of The Group: that they are usually, if not always, expected to be dominant. This means rarely getting a chance to occupy the center of attention as an enthusiastic bottom. The fact that Max is bisexual and met an eager top gave him an opportunity to be submissive in a way that's obviously not readily available for most straight men—except through fantasy.

Unlike Max's girlfriend, some women thoroughly enjoy opportunities to dominate men sexually. But a number of men have told me that even though they fantasize about being dominated, they have a very difficult time actually surrendering to women—especially those they care about. Perhaps this is one reason that so many prostitutes report that male customers who make specific requests usually want to be dominated in one way or another.

Max emphasizes a connection between physical size, strength, and sexual power roles. He's not alone in this. In most people's minds, the larger, stronger person, regardless of gender or sexual orientation, is expected to play the role of top. Even gay men complain that their options are sometimes limited by their size. Large, muscular gay men say other men usually want to cast them in the dominant role, even when they would prefer a more receptive position.

SYMBOLS OF POWER

Erotic interchanges energized by the dynamics of power often cast the protagonists in clearly hierarchical roles. In stories told by The Group, actual social relationships containing power discrepancies—such as teacher-student, parent-child, older sibling–younger sibling, doctor-patient, cop-suspect, master-slave, prisoner-guard, boss-employee—often provide the framework for memorable turn-ons.

Unfortunately, any social role with a built-in power inequity also can be used for sexual exploitation and manipulation. Consequently, society makes at least halfhearted attempts to restrict sexual contact in such situations. But these restrictions often backfire by increasing the erotic tension within these dyads, especially when power inequities are combined with the naughtiness factor. Holly's favorite fantasy relies on this tantalizing mix to create a wild yet self-affirming turn-on:

> I'm driving home from work, speeding down a twisting, windy road in my sports car. Ahead of me I see a police car hidden in the bushes. I decide *not* to slow down. As I fly past, his siren wails as he takes off after me. I pull over and watch in my rearview mirror as a gorgeous, dark-haired man gets out of the patrol car. He asks me if I realize I was speeding. "Of course," I tell him. "I did it on purpose so I could meet you."
>
> He says that instead of writing me a ticket, there is another way to take care of it. I get into his patrol car and we go find a secluded place. In the backseat, we begin kissing and undressing each other. I climb on top of him and ride him for what seems like forever. Afterward he drives me back to my car and asks to see me again. He says I am the most exciting woman he has been with. I tell him that maybe he'll be lucky and stop me again sometime, as I drive off without a ticket!

When Holly describes what makes this fantasy so exciting for her, she raises some interesting points:

> First there's the pure danger of fucking a police officer in a patrol car. But mainly it's the fact that cops are so manipulative and power-hungry, but here I manipulate one of them

> with my body. I feel control over him. The most intense part of the fantasy is when I drive away without a ticket. I have triumphed!

Once again we see the paradoxes of power. Holly conjures up the image of a power-hungry cop so that she might prevail over him. Erotic scenarios fueled by the dynamics of power often involve similar reversals. The climax of the scene always occurs, as it does for Holly, when the dominator becomes the dominated, the aggressor submits.

ESTABLISHING SAFETY

More than a few people have felt an erotic pull toward someone whose social role, race, financial resources, or age created an imbalance of power. Such attractions are usually best explored in fantasy or in role playing with a consenting partner. Even sexual situations that would be abhorrent and traumatic if they actually happened—rape or incest, for example—can become harmless sources of excitement in fantasy or playacting. But what about people who are excited by actually manipulating, overpowering, and violating those who are helpless and vulnerable? Such people do exist and understandably create suspicion and fear, especially among women.

I'm struck by how frequently women members of The Group describe the men in their wildest encounters as either gentle or "strong, yet gentle," while relatively few men emphasize the gentleness of women. I believe a collective female awareness of the potential for unwanted male aggression lies behind this focus on gentleness. For many women, if a partner is gentle he automatically becomes safer. When security is assured, women can appreciate opportunities for healthy dominance and powerful surrender just as much as men.

CORNERSTONE 4: OVERCOMING AMBIVALENCE

Whenever I speak to groups about the first three cornerstones, nodding heads, knowing grins, and insightful comments communicate a sense of recognition and understanding. After all, if people didn't immediately recognize the cornerstones, I'd have to question seriously my belief that they provide the most common excitement-boosting obstacles. But as the discussion moves to the erotic significance of ambivalence, signs of recognition are initially outnumbered by puzzled looks. A common question is: How could wanting and not wanting, liking and not liking, being drawn toward and being repulsed, be anything other than confusing and inhibiting?

People begin to understand how overcoming ambivalence can be a sexual intensifier when they view the experience of mixed feelings in a larger context—as an inevitable aspect of the human condition. From birth we are all compelled by an inner urge to engage in life with curiosity and wonder. At the same time, the more we discover about the realities of existence, the more we realize that life is painful, dangerous, unpredictable, incomprehensible, notoriously unfair, and then, as the joke goes, we die.

Among life's harshest realities is the fact that virtually everyone is hurt by love, beginning in our family relationships. Even those who are fortunate enough to feel basically loved by one or both parents still must cope with being misunderstood or ignored at times. We are so dependent as young children that it's difficult to imagine *not* being emotionally wounded by the people upon whom we depend completely for nurturance and love.

As we grow, our dependency decreases, and with it some of our vulnerability. But we continue to long for human connection despite the inevitable hurt. The need to reach out versus the imperative of self-protection is such a fundamentally human conflict that it affects all areas of life, including our eroticism.

FROM AMBIVALENCE TO PASSION

It is helpful to think of ambivalence as an internal form of the erotic equation. When someone is sexually ambivalent, the two key ingredients for high excitement—attraction and obstacles—are both active within the same person. In the right proportions, under the right circumstances, the result can be a compelling turn-on.

Most people don't readily think of ambivalence as an aphrodisiac because it is most likely to be so just at the moment when it disappears. After all, it isn't ambivalence alone that turns people on but rather the transformation of mixed feelings into a single-minded focus on pleasure. For this reason, when ambivalence adds intensity to sex, it usually operates in the background. By the time we are highly aroused, ambivalence has at least momentarily been swept aside.

This process is easy to see when ambivalence plays a part in The Group's peak encounters. You can't help noticing how reticence yields to passion as Lydia, age twenty-seven, resolves an ongoing struggle between fascination and fear:

> My boyfriend kept asking me to try anal sex but I always refused. Yes I was curious, but I just wasn't sure if it was the right thing. Besides, my friends *all* told me it was very painful and I believed them. Eventually, I decided to try, but I couldn't help wondering if I was making a big mistake.
>
> We used salad oil as a lubricant. I liked the way it felt as my boyfriend relaxed me with his finger. For the first five minutes I was too scared to get into it. Then I said, "Why don't we try it with me on top?" Having just a little more control over my position did the trick for me. What a pleasure it was to try something new. Feeling his penis sliding deep inside of me was wonderful. Anal sex is now one of my favorite things. How wrong my friends were. I've never felt any pain at all.

Although Lydia seems aware that her initial reluctance contributed to her arousal, it would not be surprising if she had left out of her story the fact of her reticence (background), concentrating instead on the joy of her new discovery (foreground). Perhaps this is one reason that only about 10 percent of The Group's peak encounters and even fewer of their favorite fantasies include explicit references to ambivalence. In discussions with clients in therapy I've noticed how they often downplay or "forget" about the ambivalence that preceded an exciting erotic event, unless I specifically ask them how they were feeling.

AMBIVALENT ATTRACTIONS

Strong attractions, whether lusty or limerent, are usually single-mindedly definite. The fact that our attractions can be so compelling, yet not controlled by logic, means that, sooner or later, most of us will come across a person who magnetically draws us while simultaneously repells us in some way. As strange as it may seem, an ambivalent attraction can, all by itself, make the object more exciting.

Those who are sexually attracted to men seem to be particularly inclined to find themselves simultaneously drawn and repelled. Among The Group, mixed feelings toward partners are mentioned much more frequently by the three subgroups drawn to men: bisexual women (25 percent), gay men (18 percent), and straight women (15 percent). Like Laura, a successful stockbroker, they usually mention traditionally masculine qualities that are both arousing *and* distasteful:

> There was a big, muscular hunk in my office who was always putting the make on me. His attitudes about almost everything disgusted me, even the way he propositioned me was so tasteless I *had* to refuse. But just thinking about him made my blood boil.
>
> Once after an office party, I let him drive me home. We made out in the car. Unfortunately, he was a terrific kisser. I invited him in. Rarely have I felt so excited. In bed he was aggressive, yet totally aware of what I wanted. His body was even better naked.
>
> I've been refusing him ever since. He's still a pig at the office but I'll always enjoy that memory.

Why would anyone be moved by such profound, erotic stirrings toward someone so distasteful? Laura is quite articulate about her dilemma:

> I can't tell you how much I resent that masculine superiority shit. I guess he gets to me because he's the exact opposite of the way I think people should be. It pisses me off to think that this tension could excite me so (I would never admit it

to anyone). The truth is, I'm incapable of feeling indifferent toward him and the bastard knows it, too.

Ambivalent attractions refuse to be limited by logic or politics, a fact that Laura reluctantly acknowledges because she's too smart not to. She realizes that the contrasts between her and "the hunk," intensified by her negative emotions toward him—not to mention his terrific body—all combine to produce an unavoidable attraction. As is so often the case, the more she tries to resist it, the stronger it becomes.

OVERCOMING AMBIVALENCE THROUGH FANTASY

Even though her wayward attraction bothered her, at least Laura enjoyed her encounter. For many others, ambivalent turn-ons are as distressingly negative as they are compelling. In such instances, the erotic mind displays an uncanny ability to convert negative real-life experiences into exciting fantasies. Notice how George, a gay man approaching forty, transforms a traumatic encounter into something positive:

> I was attracted to a large football player type who was the bouncer at a gay disco. I eventually went on a "date" with him which just meant going to his house for sex. The man turned me on no end, even though the more we talked the more I realized he was a pig-headed jerk. He said he had fathered a couple of children and that women loved to be treated like dirt.
>
> But the most unpleasant part was that I was going to get fucked, which I normally would enjoy, except for the size of his penis. Just as they say in dirty magazines—it was the dick of death! Anyway, I suggested some other act but he said he was "sick and tired of hearing this shit from faggots." So he pinned me down and forcibly fucked me. I'm not sure if he used a lubricant, but the pain was horrible. I lost my erection and prayed he would finish as quickly as possible. Because, I believe, my resistance was a turn-on to him, he *did* come quickly.

> Afterward I felt dazed and, amazingly, I was almost affectionate to him as I left, saying something like "I'll see you soon." Only later did I realize I had been raped! I would *not* like to repeat this experience, but even now I sometimes think about it while masturbating. In my fantasy, the pain doesn't really hurt. But that jerk can still turn me on.

Notice how ambivalence combines with intense power dynamics to make this encounter/fantasy memorable in spite of (or because of?) George's distasteful partner. In a roundabout way, this experience is also energized by a deep longing for the sensitivity and caring that are so noticeably absent. George explains:

> Even as a kid I admired supermasculine men, the ones who never had to worry about being called a sissy. I remember imagining that one of them—the rougher and tougher the better—would fuck me with love and respect. I knew it would never happen, but I guess that's what fantasies are for.

It's fascinating how George is able to retain in fantasy the exciting features of the encounter while filtering out the distasteful, hurtful parts. Both men and women report using this technique.

THE AMBIVALENCE OF LOVING

Sometimes the drama of overcoming ambivalence is most poignant in those on-again, off-again relationships that can be so tempestuous. Notice, for example, how a deep reluctance joins forces with longing for Joyce, a woman who was divorcing her husband. She hadn't seen him in four months, or had sex with him for even longer:

> He called me at work to say he needed to see me. I was hesitant because in the past these encounters have led to either fights or outrageous sex. At the time I didn't want either from him. Yet he was very persuasive so I agreed to meet him in spite of my better judgment.
>
> From the moment I saw him, it was like the beginning of our relationship all over again. Sexual sparks were flying

everywhere. He knew the right places to touch me and the perfect words to say. And he used all his tricks until I was like jelly. Incredible!

Joyce's explosive encounter is defined and energized by the push-pull of ambivalence. Her desire to avoid him only intensifies the magnetism of his "tricks." Yet by the end of the story, her ambivalence is nowhere to be seen. In a burst of passion, ambivalence is transformed.

THE CORNERSTONES IN ACTION

None of the cornerstones is *required* for sexual arousal. A strong mutual attraction combined with a vital sensuality can, by themselves, create a very satisfying turn-on. But as you have seen, the cornerstones are extremely effective arousal intensifiers. And because excitement is notably heightened in the peak moment, all the features that contribute to our arousal, including any of the cornerstones, are especially visible.

You've probably noticed that many of The Group's encounters and fantasies include more than one cornerstone—even though I've deliberately selected stories that are relatively pure examples of whichever cornerstone I'm discussing at the time. Three-quarters of The Group's memorable encounters and fantasies contain at least one cornerstone, and about 40 percent mention two or more. Zack alludes to all cornerstones except ambivalence:

There was this girl that I wanted for a year and had often used her as a model during masturbation sessions [*longing*]. When we finally had sex for the first time it was great. I enjoyed being the aggressor, since I had always been the passive one in my previous sexual relationships. I enjoyed having her submit to me and let me do as I pleased [*power*]. What really turned me on was seeing her naked and hearing her breathe deeply. We were also in a place that was risky to be fooling around in [*naughtiness factor*]. I had just about come by the time I had her clothes off. It was extremely arousing when she started touching me. I had always imagined what it would be like and it turned out to be even better.

Many people have a particular affinity for just one or two of the cornerstones, while the others are of little interest. In general, those cornerstones that were most consistently a part of your earliest experiences of arousal are likely to be the ones you respond to today.

Sometimes, although not always, it is essential to become aware of which cornerstone or cornerstones excite you. I learned this when Alice entered therapy with me because she was tired of acquiescing to sex with Hugh, her husband of nineteen years. Rarely had she felt genuine desire during her marriage. But now an undeniable revulsion was forcing her to stop going through the motions and discover why she was so turned off.

The reasons for her dilemma quickly became apparent as she described how much she had enjoyed sex with Hugh before they married. Both were active in the church youth group. But Alice was a bit wilder and enjoyed seducing Hugh, who, although horny, believed in abstaining from sex until marriage because he hoped to become a minister. When I explained the naughtiness factor, Alice understood it immediately and soon realized what had gone wrong with her sex life. "The moment we married," Alice proclaimed, "I felt completely different about sex. Now it was proper, a duty, *a bore*!" Her challenge was to restore a little naughtiness to her relationship, not an easy task for a minister's wife.

As you zero in on the cornerstones that are most important to you erotically, think of each one as operating on a continuum, ranging from subtle to dramatic. Many of The Group's stories have considerable drama. But keep in mind that less can be more. Sometimes just a hint of naughtiness, a tease of anticipation, or a whisper of domination is the right amount.

Not only is there tremendous variation in the *intensity* with which each cornerstone comes into play, but *timing* is also important. Both longing and ambivalence usually create erotic tension *preceding* sex. The first passionate embrace may actually cause a reduction in longing or ambivalence—accompanied by an explosion of excitement. On the other hand, violating prohibitions and searching for power are often most exciting *during* an encounter. Sometimes a cornerstone fuels arousal throughout an entire encounter or fantasy. But the effect of any cornerstone may also come and go in an instant when, for example, a fleeting thought of someone watching boosts excitement with a short-lived burst of naughtiness.

Keep in mind that you may not always be aware of the things that excite you. Sometimes a cornerstone works on the edge of consciousness—a subtle impulse you don't have to, and may not want to, think about. Sometimes awareness actually gets in the way, especially if you are being excited in ways you wish you weren't. Then consciousness turns to self-consciousness; the spell is broken. But in some cases awareness seems to be crucial for full enjoyment of the cornerstone. After all, if you're not aware of *feeling* naughty, how can you possibly enjoy *being* naughty? What's the point of *un*consciously longing for someone? Why bother surrendering if no one notices?

Observing the effects of the four cornerstones reminds us again that intense eroticism is paradoxical and unpredictable. Almost anything that arouses us may—under different circumstances, or with greater or lesser intensity—also turn us off. And virtually anything that inhibits us sexually can reappear later as a turn-on. Once we grasp the implications of this, we are able to appreciate more fully the richness and complexity of our erotic minds. With this deepened appreciation, we can enlarge our perspective further still by considering some of the most ancient and powerful of all aphrodisiacs.

4

EMOTIONAL APHRODISIACS

Feelings are potent sexual intensifiers—but not always the ones you expect.

In a far corner of the globe, hidden from view, is a magical substance capable of evoking sexual desire in anyone lucky enough to find it—or so holds one of the oldest and most persistent of all myths. The perennial search for a "true" aphrodisiac has motivated people to sample everything from barks and roots of trees, leaves and flowers of exotic plants, innumerable animal extracts and body parts, to a host of concocted potions. More recently scientists have taken an interest in the biochemistry of attraction and arousal. But despite our best efforts, a reliable sex-enhancing substance remains little more than a dream.

Aphrodisiac quests have lured seekers for centuries. But few have given much attention to the readily available and remarkably potent aphrodisiacs much closer to home—our emotions. Perhaps emotions are so much a part of everyday life that we take them for granted, erroneously believing that anything as ordinary as a feeling couldn't possibly make the difference between ho-hum sex and sex that moves and satisfies us profoundly. Yet emotion plays an enormously important role in sexual desire, arousal, and fulfillment.

Whereas the four cornerstones we explored in the last chapter are the building blocks of eroticism, emotions are the energizers. Feelings make sex matter. True, some purely lusty encounters and fantasies appear to be practically emotionless.[1] Although sex with minimal emotion can offer a pleasurable distraction and a welcome release of tension—while keeping the participants shielded from vulnerabilities—such experiences are limited in the amount of satisfaction they can bring. Among The Group's peak turn-ons, not one single encounter or fantasy—not even anonymous ones with strangers—is free of emotion. Without emotion, we simply cannot have a peak.

Emotion has always been a focus of modern psychology. Yet no matter how much feelings are analyzed, there is very little consensus about what they actually are, where they come from, how they work, or what they mean. No one can even agree on how many different emotions we're capable of having. William James, the great psychologist, speculated that "there is no limit to the number of possible emotions which may exist."[2]

Disagreements aside, just about everyone recognizes the crucial role of emotions. German philosopher Nicolai Hartmann was unequivocal: "emotions are the stuff of life's inner content and the basis of its richness."[3] Reaching a similar conclusion after an exhaustive study of emotion in psychology, James Hillman proclaimed that "emotion is the essence of life."[4]

Whenever The Group describes peak encounters or favorite fantasies, their stories are animated by a wide range of emotions. Feelings associated with peak eroticism tend to be of six distinct types, listed here starting with the ones mentioned most frequently:

Exuberance, including joy, celebration, surprise, freedom, euphoria, and pride.

Satisfaction, including contentment, happiness, relaxation, and security.

Closeness, including love, tenderness, affection, connection, unity (oneness), and appreciation.

Anxiety, including fear, vulnerability, weakness, worry, and nervousness.

Guilt, including remorse, naughtiness, dirtiness, and shame.

Anger, including hostility, contempt, hatred, resentment, and revenge.

The first two sets of emotions—exuberance and satisfaction—are so universally rated as highly important aspects of peak arousal that we must conclude that peak turn-ons, by definition, generate these unabashedly positive feelings. However, as crucial as these terrific feelings are to sexual enjoyment, it's not quite accurate to think of them as emotional aphrodisiacs because they do not produce or intensify arousal so much as they follow from it. Therefore, I call exuberance and satisfaction the *response emotions* because they are the rewards of high arousal, not the cause.

In this chapter our focus is on the other four sets of emotions—closeness, anxiety, guilt, and anger—all of which have unmistakable aphrodisiac powers. As you consider the role of emotion in your own peak turn-ons, you probably won't have any trouble identifying the presence of the "positive" ones. After all, emotions such as love, tenderness, and affection are most likely to fit your ideals about the way sex should be. "Negative" emotions such as anxiety, guilt, and anger probably do not spring to mind nearly so readily when you contemplate peak turn-ons. Yet on certain occasions negative feelings energize your eroticism at least as much as love and affection—although you may not always recognize them. I call these feelings the *unexpected aphrodisiacs.* Just as surely as anxiety, guilt, or anger can disrupt sexual enjoyment, they can also enhance it. Thus they are the paradoxical emotions.

Look closely at your peak experiences and you'll discover that one or more unexpected aphrodisiacs contributed to your excitation, most likely in a subtle fashion that easily can be ignored. If you do become aware that negative emotions sometimes add dynamism and drama to your turn-ons, you may understandably be surprised or disturbed. Rest assured, however, that any risks you take in examining the place of the unexpected aphrodisiacs in your erotic life will be richly rewarded. The more you're able to recognize the unexpected aphrodisiacs, the greater will be your comfort with the full range of emotions associated with passionate sex.

In the realm of the erotic, all your emotions are part of a larger whole in which the concepts of "positive" and "negative" are often meaningless. As we explore closeness, anxiety, guilt, and anger you will see that feelings are fluid rather than static. All emotions—whether they initially appear positive or negative—can be transformed from one into another. Even anger, usually considered the very antithesis of caring and tenderness, can be metamorphosed by your erotic mind into appreciation and love.

THE SEARCH FOR CLOSENESS

Of all the impulses that motivate erotic adventures and daydreams, even those that appear to be completely impersonal, none is more fundamental than the urge to engage with another human being—if only for a moment. When sexologist and sociologist Dr. Ira Reiss set out to identify features that all sexual experiences have in common—regardless of historical setting or culture—he came up with only two: physical pleasure and self-disclosure.[5] His insight is especially relevant to peak eroticism, during which ultra-personal desires are played out in relatively unedited form. To share your sexual quirks and eccentricities with another person is a bold act of self-revelation. No wonder you feel profoundly validated when your partner revels in your self-expression.

Perhaps it's self-disclosure combined with positive responses from their valued partners that cause virtually all members of The Group to recognize one or more of the closeness emotions either before, during, or after memorable sex: love, tenderness, affection, connection, unity (oneness), admiration, and appreciation.

The urge to connect has the most dramatic aphrodisiac effects during limerence, when cravings to be close are so overpowering that the lovers merge into a joyous oneness. Particularly when romance is new, feelings of love and sexual enthusiasm are synergistic: each intensifies the other. For those in limerence, love is unquestionably the best aphrodisiac and may even seem to be the only one. Whenever The Group describes romantic peaks, the closeness emotions always appear. How could it be otherwise?

For many people the desire to create a sense of connection even with a casual or anonymous partner is an erotic motivator. Of course, there is the desire to have skin to skin contact, to feel the weight and heat, to hear the sounds, and to savor the sight of an aroused other—all of which bring the participants into the closest possible physical proximity. Although some encounters involve no expressions of affection, more commonly the protagonists *act* as if they're close—kissing, holding, caressing—at least part of the time, even if they know nothing about each other. Even for those who can remain emotionally cool when they're sexually hot, the search for closeness is still woven into the fabric of their erotic motivation, no matter how fine or carefully concealed the thread.

CLOSENESS AND THE FOUR CORNERSTONES

As you saw in the last chapter, the four cornerstones of eroticism become intertwined with arousal because they are universal human experiences that involve overcoming obstacles. Sometimes these existential barriers heighten excitement by stimulating your desire for closeness. You've no doubt noticed this effect if you've ever longed for someone who wasn't readily available. Chances are your desire to be close increased in direct proportion to the difficulty of getting together. Similarly, if you've violated prohibitions by becoming involved with someone your friends, your family, or even you yourself considered inappropriate, you may have noticed that social disapproval made you cling more tightly to each other than you would have otherwise.

Frederico, a member of The Group in his early thirties, describes his struggle to spend more time with his bisexual girlfriend. His story is a perfect example of the dynamic relationship among the four cornerstones and the search for connection:

> About a year ago I started dating again after being dumped by a lady I lived with for four years. She told me I didn't have a clue about being intimate and constantly demanded that I tell her how I was feeling, which I thought I did. But it was never enough. After a while I gave up.
>
> I was pretty much convinced I didn't know how to love a woman when I met Audrey at a friend's party. Our flirting made me feel sexual again. The best part was that she was so *friendly.* That first night we stayed up until the wee hours getting to know each other and making out on her

couch. She told me she had a girlfriend but also wanted a boyfriend. I admired her honesty. And since I enjoy pictures of two women getting it on I thought, hey, she's just the wild type. Besides, her openness made me feel open too.

I liked her more each time I saw her. One night Audrey introduced me to her girlfriend, which was more awkward than I thought it would be even though she seemed nice. Afterward Audrey and I had the hottest sex I've ever had in my life. We couldn't get enough of each other. I wanted to turn her on more than her girlfriend. I even imagined that her girlfriend was spying on us through a crack in the curtains. The thing that blew me away was how much I wanted to be close to her, to tell her *everything*, all the stuff my old girlfriend tried to drag out of me was spilling out to Audrey and I was loving it.

These days it still pisses me off that *she* decides how often we see each other (about two or three times a week). Most of my friends tell me I'm wasting my time, but they don't understand what it's like to be with her. Yes, I've considered cutting things off, but I can't imagine feeling so good with anyone else. What can I say? I'm in love.

Frederico's desire to be close to Audrey is enhanced because he must share her affection with someone else. The fact that he wants more time with her than he can get assures that he has plenty of opportunities for longing. In addition, Frederico's reference to his friends' advice suggests that their disapproval strengthens his attachment, underscoring the sense of "us against the world" that energizes so many unorthodox relationships.

In passing, Frederico also touches on the role of the two remaining cornerstones in boosting his affection for Audrey: searching for power and overcoming ambivalence. He doesn't like the fact that she controls how often they see each other. Yet her position of power keeps him, quite literally, in hot pursuit. More often than not, the desire to be close is felt most keenly by whichever partner is less secure—Frederico in this case. Nor is it unusual for someone in Frederico's position to have bouts of ambivalence about the relationship. He naturally wants to avoid being hurt again if this affair is doomed, yet each time he reaffirms that Audrey is worth the risks, his ambivalence is overpowered by his need to be close.

CLOSENESS AS AN ANTIAPHRODISIAC

Hardly anyone needs to be convinced that feeling close to someone can be a turn-on. Yet it's equally important to realize two ways that emotional connections can dampen rather than stimulate desire: (1) when closeness becomes an obligation or demand and (2) when it threatens to dissolve the separateness that is the basis of all attraction.

In Frederico's story, it's impossible to ignore the contrast between his role as a closed, nonintimate male with his old girlfriend and his eagerness for total involvement with Audrey. We know practically nothing about his old relationship. But in his own analysis of why he felt so much closer to Audrey, Frederico writes, "Nancy [his old girlfriend] made me feel like intimacy was a chore—something to get out of. I also felt completely inadequate to satisfy her. Proving to Nancy that I loved her had become a test I was destined to fail."

Of this I am sure: whenever closeness feels like a requirement—something owed rather than inherently gratifying—it inevitably switches from an aphrodisiac to an antiaphrodisiac. The erotic mind may enthusiastically gravitate toward the risks of intimate self-disclosure. But once you become convinced that you cannot meet that challenge, your enthusiasm changes into avoidance. Many long-term partners set each other up for a similar fate by allowing their closeness to become a "should" rather than a choice.

Even couples who manage to avoid making intimacy an obligation will eventually face its paradoxical nature. In early romance the urge to merge magnetically draws the lovers to each other. Yet once they are doing everything together, developing feelings and opinions as a unit rather than as two individuals, they undermine the sense of otherness that was the original basis of their mutual appeal.

ANXIOUS APHRODISIACS

As an organism determined to survive, you are equipped to assess risks and avoid danger. Fear makes your heart and breathing rates increase. Yet each breath becomes shallower as your chest muscles and diaphragm tense up in a protective armor. Your arms and legs prepare to defend, hit, kick, or escape. Your vision, hearing, and sense of smell become ultra-sensitized. Yet your sense of touch tends to numb so that you won't be distracted by pain. You're pumping out adrenaline to optimize your physical strength and alertness.

Whereas fear is usually a response to fairly specific and imminent dangers—including all kinds of physical and emotional attacks—anxiety is concerned with *potential* dangers, whether conscious or unconscious. Research at the Menninger Foundation has shown that the unconscious mind does not distinguish between real and imagined threats.[6] Even when the focus of anxiety is "all in your head," your body responds as if the danger is real. The subjective experiences of fear and anxiety have so much in common that I use the two words interchangeably throughout our discussion.

The relationship between anxiety and eroticism is intricate and paradoxical. If you are highly anxious in a sexual situation, your physical capacities for arousal or orgasm or both will usually be short-circuited. Modern sex therapy can effectively teach sexually anxious people how to reduce fear and create opportunities for pleasure. However, to view anxiety solely as antithetical to arousal is to blind ourselves to a richer and more challenging reality: just as surely as anxiety can disrupt arousal, it can also create, focus, and intensify it. Depending on the situation and the individuals involved, anxiety is either an antiaphrodisiac or an aphrodisiac—occasionally both.

Anxiety intensifies arousal by contributing to a generalized state of physical excitation. All forms of excitement, sexual and nonsexual alike, increase muscular tension, blood flow, and heart and breathing rates. Consequently, your body responds similarly to anxiety-provoking and sexually arousing situations.

For instance, some men and women spontaneously experience sexual arousal in a wide range of frightening situations, including everything from fights to roller-coaster rides to sexual assaults.[7]

ANXIETY AND THE FOUR CORNERSTONES

In addition to the physiological links between fear and sexual excitement, anxiety is a natural response to many of the risks and dangers inherent in the four cornerstones. During the experience of longing and anticipation, the urge to be with the one you desire is paramount. But within your yearning may lurk uncertainty and fear. Is the attraction mutual? Will you ever see her again? Will he still feel the same way about you? Does she miss you as much as you miss her? Will the two of you be as sexually compatible as you hope, or as you have been before? Are you setting yourself up for hurt or disappointment?

Longing—and the associated anxiety—reaches its zenith in romantic love, for it is here that your deepest vulnerabilities are exposed. In Dr. Tennov's study of limerence, the most commonly reported physical reactions to falling in love were heart palpitations, trembling, pallor, flushing, and general weakness—all the common fear responses.[8] The great French philosopher of romance Stendhal captured the interconnection between romance and anxiety when he concluded that "the pleasures of love are always in proportion to the fear."[9]

Both lusty and limerent attractions seek a response from the other. Consequently, they both leave you open to potential rejection, a prospect that grows more frightening in direct proportion to your level of attraction and need. No wonder we often act like nervous adolescents in the presence of those who turn us on most. Quite naturally, we experience our nervousness as an annoyance and embarrassment and try to hide it. But Dr. Tennov insists:

> However unappealing it may be in a universe conceived as orderly and humane, the fact is undeniable: fear of rejection may cause pain, but it also enhances desire.[10]

Anxiety can also function as an aphrodisiac when it is a response to the risks involved with searching for power. When you are a top in a dominance-submission scenario, the rush of feeling powerful may be tempered by the weight of responsibility for the outcome. Conversely, when you submit to the power of your partner, you may savor your role as the center of attention, but you may also fear being overwhelmed and pushed somewhere you do not wish to go.

Anxiety has an especially close connection with violating prohibitions. Although you may not always be aware of it, at least a portion of the excitement generated by sexual rule-breaking results from the fear of being caught, disapproved of, or possibly even punished. Ron, a forty-year-old member of The Group, remembers an encounter from fifteen years ago as if it occurred yesterday. Notice how his overwhelming anxiety miraculously translates into heart-pounding passion:

> One of the most memorable sexual experiences of my life took place while I was living and working in a London public house. I was having an affair with the owner's wife. One day I went down to the cellar to change a barrel of beer and she just happened to be there too. Even though the pub was busy, we started fooling around. Without speaking she grabbed at the obvious bulge in my crotch as I groped my way inside her soaking wet panties. I felt her whole body tremble and I was shaking too. She begged me to stop because someone might hear us from the kitchen but she didn't miss a beat as she unzipped my pants. I never knew my dick could get so hard! Frantically trying to hide, we moved on top of the barrels, but it was so cold she let out a scream (which I squelched by kissing her).
>
> Eventually, we huddled under the stairs where we could hear the door open. This would give us a few seconds to pull apart and straighten up. We might look suspicious as hell but at least we wouldn't be caught red-handed. We were both scared shitless, but I've never felt more alive in my life. She sucked me off with such force that I exploded within a few minutes. I was still so hot I made a dive for her pussy to lap up her juices. What made this encounter doubly exciting was the fact that her husband was directly above us pulling

pints for customers. It took me days to calm down from that one!

Ron's tale is a bundle of contradictions. It certainly illustrates the two top ways that adults boost their arousal by reveling in the forbidden: (1) by choosing an inappropriate partner and (2) by having sex in a place where risk of discovery is high. But his story also raises an obvious question: How can anybody function sexually in the face of so much fear? Ron muses about this himself:

> I've heard of mountain climbers and race drivers getting off on an "adrenaline high." But that's not me at all. I don't like to be scared and definitely not in sex. I remember a couple of times when I couldn't get it up just because my girlfriend's roommate was in the next bedroom. And here I was practically asking to get my ass kicked, or maybe even killed, and I became an unstoppable sex machine. My own fear and seeing the fear and desire in Martha's face made me wild. I didn't care about the danger—at least not until later.

Ron's on to something important here. Normally, if anxiety is going to have an aphrodisiac effect it will usually be at low to moderate levels. Too much of it quickly becomes unpleasant and disruptive. In his fascinating book *The Dangerous Edge,* Michael Apter explains that we are most likely to experience high excitement on the razor's edge between safety and danger when we know the risks but take comfort in the belief that we are somehow protected from catastrophe. Yet in some situations of super-intense arousal such as Ron's, dangers that would normally make us retreat have the opposite effect: they spur us on to greater levels of risk-taking. Apter concludes:

> In the excitement-seeking state, one can get to the dangerous edge with impunity. And one does indeed often try to get as close to the edge as possible in this state, because that is where arousal is likely to be at its highest.[11]

Here's one more factor to consider: if anxiety is interwoven with high arousal right from the start, as it was for Ron, the two can increase in tandem. On the other hand, if strong anxiety takes hold *before* arousal is established—as, for example, when a man worries all day about whether he'll be able to get an erection that night—his fear responses are likely to choke arousal before it begins.

If you examine the role of anxiety in your own peak encounters, you're most likely to notice a relatively small amount of it—at least compared to Ron. Reread any of the stories of naughty sex in the last chapter and you'll notice a manageable hint of anxiety that can amplify desire. Remember the young woman who defiantly reveled in sex while parked in their driveway with the boyfriend her parents hated? Or the couple who briefly escaped their kids for an exciting encounter in a park, their arousal enhanced by looking over their shoulders? Their risk-taking was more fun than truly dangerous.

In the process of overcoming ambivalence, the special alchemy of desire and fear reveals itself with particular clarity. Fear plays at least some role in the wanting/not wanting of most ambivalent attractions. You can feel the conflicting impulses in this story told by Ben, a thirty-eight-year-old married engineer who still recalls an especially exciting, yet uncomfortable, encounter with his first girlfriend almost twenty years earlier:

> Meg was an incredibly attractive girl who was a complete mismatch for me. She was rather kinky sexually while I'm more of a high-tech bookworm type who is pretty timid in bed. Meg was always pushing me into dangerous territory and I never knew what to expect. Since I was looking for a serious relationship I had told her that we had to stop having sex and I just wanted to be friends.
>
> That evening a few friends were due to arrive for dinner when she started playing with me. All of a sudden she unbuckled my belt and rubbed her tits all over my face. "Are you *crazy!*" I whispered. She said, "Yes! Now shut up and relax!" I made a couple of halfhearted attempts to reason with her but I was getting so hard especially when she stuck a hard nipple into my mouth.
>
> I panicked when I thought I heard a car in the driveway. As I struggled to stay calm, she went down on me and used an incredible undulating action with her throat on the head of my cock. Even as I was about to explode, images raced

> through my mind of the doorbell ringing just as I was coming. In a few minutes I closed my eyes and gave in to an incredibly intense orgasm.
>
> No sooner had I come when the bell did ring. I stumbled to the bathroom where it must have taken ten minutes for my hard-on to go down and my heart to stop racing. When I finally came out she said, "Where ya been, baby?" with an impish grin. Our relationship continued for another six months before, out of desperation, I cut it off completely.

Ambivalence, accompanied by fear, is the central theme in Ben's encounter. But like most peak turn-ons, Ben's has many different layers. Notice, for instance, how Meg's aggressiveness interacts with his seemingly reluctant submissiveness to add yet another dimension to their arousal. You can also see the arousal-enhancing dynamics of the naughtiness factor expressed both in Ben's anxious images of getting caught and his focus on the contrast between Meg's "kinkiness" and his own conventionality.

FEAR TAKES A HOLIDAY

Of all the forms that fear can assume in erotic life, one of the most wonderful occurs when it yields to an overarching desire. Sometimes that moment has a deeper significance and can mark a turning point, the opening of a door to a new and enlarged sense of self. Notice how Glynis, age forty-four, describes the first time she made love with another woman:

> She was twelve years older than I—my professor in grad school. I had invited her to stay at my apartment while hers was being painted. I had no intention of anything happening, although I must admit I was in love with her. We had several nights of holding hands and talking all night. On the third night, we were making dinner (I had a salad in my hands) when she pushed me against the yellow refrigerator. I couldn't believe this was happening. For years I had feared that if I ever made love with a woman I would *dissolve*—or the other person would *die*—yet here I was feeling incredibly *alive and real!* All my years of fear went out the window. I knew who I was and I loved it.

> Our lovemaking went on for several days. By far the most arousing aspect was the permission I felt to express deep feelings for someone. I felt free and joyous. Also profound for me was the fact that through this experience I had come to love myself as a woman. Prior to that day I think I had wished in some subtle way that I was a man. This was a religious experience for me—a feeling that everything was divinely perfect.

Glynis doesn't tell us if this watershed encounter put an end to her struggles with being a lesbian. I would guess not. Most lesbians and gay men I know report that coming out is not a one-time event but an ongoing process in which fear, reluctance, and often guilt are all overcome by desire, only to return later—sometimes immediately after orgasm. Until the development of a deep self-acceptance counteracts the last vestiges of internalized homophobia—which sometimes takes a very long time—many gays describe repetitious cycles in which the triumph of desire over fear produces a special kind of intensity.

RETHINKING PERFORMANCE ANXIETY

Everyone knows that performance anxiety is a common fear that can devastate those who constantly worry about their sexual adequacy. Preoccupation with performance can cause people to shift their attention permanently from the pleasures of sex to the potentials for failure. Sex therapists, not surprisingly, are universally down on performance anxiety. But most of the time a little performance anxiety is completely harmless. And much more often than you might expect, performance anxiety can actually be an aphrodisiac.

Part of the validation most of us get from sex usually includes a feeling that we've been good lovers. We enjoy it when our partners compliment us on our sexual abilities. But inherent in that enjoyment is the knowledge that things might not have gone so well. There's no question that sexual satisfaction is greatest when pleasure—often combined with a desire for intimacy—is the primary motivation for sex. And if the urge to perform becomes too strong, sooner or later there will be trouble. Yet just a touch of performance anxiety can be part of a general atmosphere of uncertainty and risk that is often highly conducive to arousal.

GUILTY PLEASURES

By the time you were two years old you were already familiar with that nagging feeling in the pit of the stomach that you now recognize as guilt. A powerful combination of fear and self-recrimination, guilt is universally experienced as unpleasant—which is, of course, the whole point. Whether guilt first came over you in response to a wordless look of disapproval on your mother's face or a spoken criticism, it evolved into a form of self-policing, the purpose of which was to forestall the withdrawal of love and avoid abandonment—life-or-death objectives to a child.

If you were lucky enough to grow up in a reasonably healthy environment, you eventually learned to experience guilt as an emotional component of conscience. Now, as an adult, when you stray too far from the values and principles you live by, guilt nudges you back on track. A certain amount of guilt benefits you as an individual and as a member of the groups with which you affiliate. In this way, healthy guilt is a social emotion that keeps you attuned to the rules and expectations you have absorbed from your community and culture and is neither debilitating nor demeaning. Be thankful for this kind of guilt. Truly guiltless people are sociopaths—those who commit monstrous acts without a shred of discomfort.

Unfortunately, most guilt is neurotic guilt in which impossible, superhuman standards produce a harsh, merciless conscience and, in the most extreme cases, a lifetime of unrelenting self-condemnation. Neurotic guilt demands that its victims suffer for all kinds of imaginary crimes of omission or commission.[12] The most extreme form is shame, a conviction held by many people that they are fundamentally and irreparably flawed. Shame is a common result of severe emotional, physical, or sexual abuse. Whereas guilt comes from the sense of doing, thinking, or feeling bad things, shame reflects an inner certainty of *being* a bad person.

Societies and families who are uncomfortable with the sexuality of their children inevitably encourage them to feel guilty whenever they have sexual thoughts or feelings. Sexual guilt converts a child's natural curiosity about the body and its wonderful sensations into something uncomfortable. The association of guilt with sexuality either completely squelches the innocent sex play that is an important part of healthy erotic development or else drives it underground to become a dirty little secret.

Every sex therapist observes daily how guilt causes sexual dysfunctions, inhibits desire, and keeps erotic fulfillment frustratingly out of reach. To put it bluntly: guilt is second only to anxiety as the world's foremost *anti*aphrodisiac. However, the erotic equation reminds us that whatever tries to block our urges can also intensify them. Guilt can be the great disrupter of sex, but it can also be the great titillater, an obstacle that heightens attraction and desire. It's an amazing paradox: guilt as the cheerleader for sexual excitement.

GUILT AND THE FOUR CORNERSTONES

When guilt acts as an aphrodisiac, it usually does so as an emotional component of violating prohibitions. You've probably noticed that in sex, as in other areas of life, some things are so good precisely because they're so bad. We recognize this when we describe a rich dessert as "sinfully delicious." Sexologist Robert Stoller has observed: "Guilt is not the price paid for being bad but the price paid for the privilege of continuing to be bad."[13] You may recall that over a third of The Group's peak turn-ons are clearly intensified by the thrill of the forbidden—the naughtiness factor. With few exceptions, encounters and fantasies that involve inappropriate partners, forbidden acts, or the risk of discovery are energized by at least a hint of guilt.

Guilt may also add to the aphrodisiac effects of overcoming ambivalence when it helps create an atmosphere of reticence that ultimately crumbles in the face of escalating desire. Denise, now a thirty-two-year-old businesswoman, recalls a particularly exciting encounter in which she violates her own sense of right and wrong:

When I was eighteen years old I got a job in an insurance office as a secretary while taking classes at the local college. One day I was in the back room when Jerry, the owner, walked in and began to compliment me on my work. He stood next to me and placed his hand on my lower back. He kept it there for a while and then began to massage me, occasionally venturing down to stroke my ass. He did all this in the most matter-of-fact, confident way.

Since I didn't protest, he continued exploring. At one point, he took me by the shoulders, turned me toward him, and kissed me directly on the mouth with a probing tongue. Next he slowly crept his hands to my breasts, caressing and teasing my nipples with his thumbs.

Inside a little voice was saying, "This whole thing is wrong. You don't even *like* him!" But I couldn't help myself. He steered me over to a table, hoisted me up, and in one fell swoop took off my panties. Before I knew it he had his dick in hand. It was a priceless piece: the largest, hardest, smoothest, most beautiful prick I'd ever seen. He took some saliva and rubbed it on himself and then knelt down and began to slowly and delicately lick me.

With me sitting on the table, my back to the wall, he poised his cock in my direction and began to slide it in. By the time he was in all the way, I could hardly move—the thing nearly touched my heart. He began to thrust, slowly at first, but steady and rhythmic. I was out of my mind with excitement. I longed for that cock to be as deep within me as possible.

Things were definitely looking good. All I had to do was sit there, his hands firm on my breasts, his breath and tongue hot in my ear, and his fine dick ramming me with beautiful precision. I knew this would be the first of many a hot afternoon getting my brains fucked out.

Denise ignores her better judgment and forges ahead. The violation of her own values, not to mention her common sense, inevitably produces some guilt. Yet if she's seriously troubled by what happens, she's keeping that to herself. It appears as if Denise mentions the inappropriateness of her behavior only to set up the forbidden circumstances of this encounter, which, of course, also includes ambivalent feelings toward her boss. Hers is a tale of ambivalence and guilt transformed into desire, of reluctance converted to the purposes of passion, of restraint replaced with freedom. But how is guilt, the great inhibitor, so thoroughly transformed?

The first principle is that when guilt is in the forefront of a sexual situation, it almost always gets in the way. After all, the purpose of guilt is to inhibit. If, however, it can be held in the background, called up only to be overpowered by arousal, it will usually produce a sense of liberation, as it does for Denise. However, when our sexual behavior involves serious violations of our own values, after orgasm guilt will probably return with a vengeance in the form of remorse.

The cycle of attraction, guilt, excitement, remorse, attraction is what makes many illicit affairs seem irresistible. For those who place some value on sexual exclusivity, yet have an affair anyway, it is logical to assume that they proceed despite their guilt. But I believe that guilt creates a type of resistance all but impossible within a committed relationship, and thereby provides a significant portion of the erotic fuel that helps make affairs so alluring. June, a church secretary in her early fifties, offers a case in point:

> I hate to admit it, but I had my most exciting sex during a two-month affair I had with a handsome man who worked with me on a fund-raising event. Most of our meetings were spent in intimate conversation, with very little real work accomplished. We were both unhappy in our marriages and found solace in each other. At first we just hugged and kissed which for me was more than enough to bring on torrents of guilt. Many times we vowed to keep our relationship platonic. But each time we broke our vow our passion grew more reckless.
>
> In contrast to my husband who no longer found me attractive anymore (I'm not sure he ever did), "Bill" was excited by me. He was sensuous, romantic, and passionate—everything my husband wasn't. These reasons explain what I did, but for me they can never justify it. My affair was wrong but neither of us was willing to stop it.
>
> The fund-raiser was a big success. When there was no longer a reason for us to meet, I was able to cut off the affair (except for once or twice). The affair was such agony,

> I hope I never do anything like that again. But there's a happy ending. Three years ago, I somehow found the courage to get a divorce. Bill helped me realize what I was missing.

June raises an important point with her last remark. She implies that had she been unwilling to have the affair and tolerate the guilt it caused, she might have stayed in an unhappy marriage. Sometimes we have no choice but to violate the rules we set for ourselves. Perhaps part of what made June's guilty pleasures so irresistible was an unconscious realization that they would act as a catalyst for a much-needed change.

PLAYING WITH GUILT

Rather than suffering over guilt, people commonly use it deliberately and playfully to spice up sex. Let's say, for instance, that you and your lover decide to sneak away from a party to fool around behind the garage. A little nervousness about getting caught adds to the fun, as do thoughts such as, "We shouldn't be doing this," or "How sleazy can I get," or maybe even, "What would my mother think?" Perhaps you flash back to an earlier time when playing "doctor" behind a similar garage really did make you feel guilty. Your reminiscence adds a welcome titillation. For a moment you might even pretend that you are running the risk of being caught and reprimanded by a parent, minister, policeman, or some other guardian of traditional morality. Yet you also hold in the back of your mind the knowledge that you are now grown and can enjoy the anticipation of punishment while avoiding the real thing.

Whenever you deliberately introduce an element of raunchiness into an encounter or fantasy, some of your enthusiasm comes from a subtle undercurrent of guilt, perhaps recalled from an earlier time. Chances are you won't be thinking about guilt consciously because you have succeeded in neutralizing its negative, inhibiting qualities by taking control of it, claiming it as your own, and mastering it by using it for your own sexy purposes. Your disobedience becomes a demonstration that your desires are superior to the pitiful prohibitions that dare to dampen your enthusiasm. Guilt reveals its richest aphrodisiac potential when it is not forgotten, but vanquished.

ANGER AND AROUSAL

Of all the feelings that find expression in erotic life, anger and its related emotions—hostility, hatred, resentment, and revenge—are the most difficult to accept. Anger is such an unpleasant emotion that most of us wish it could be excluded from the sexual arena. And yet we are regularly confronted with the intersection of anger and sexuality in everyday language in which the word "fuck" not only means to have sex, but is also a widely used expression of hostility and mistreatment, as in "fuck you" or "I got fucked over."

Most of us think of anger as a dangerous, destructive emotion—which, of course, it most certainly can be—but its primary function is self-protection. Fear alerts you to danger, but anger helps you mobilize the energy necessary to take action. As Carol Tavris says in her fine book *Anger: The Misunderstood Emotion,* "anger is ultimately an emphatic message: *Pay attention to me. I don't like what you are doing. Restore my pride. You're in my way. Danger. Give me justice.*"[14]

Anger is self-protective, and because risks and dangers abound in the erotic adventure, we shouldn't be surprised that sex and anger are intricately linked, with results ranging from positive to destructive. Obviously, those who are chronically angry, who feel as if life itself has done them wrong, are rarely available for sensitive lovemaking, although some are very interested in hostile sex. More than a few angry people lose interest in sex altogether or are plagued by sexual dysfunctions. Angry couples can go either way. Some conflict-ridden partners rely on clashes between them to add drama to their sex lives, while others sink into sexless bickering or outright war.

Luckily, most of us are neither ruled nor defined by our anger. Instead we occasionally become angry, with varying degrees of comfort or distress. In examining your own peak turn-ons, don't be surprised if you come across one or more in which anger functioned as an aphrodisiac. The role of anger as an arousal intensifier is crystal-clear when a fight or argument is followed by a passionate reconciliation, as in this story told by Sheila, a graduate student in her mid-thirties:

> I remember a night in our second year of marriage when my husband Rick and I had a terrible fight. At least I felt it was terrible. Rick says it was no big deal. It started out as a normal argument until Rick was really getting pissed off. I felt so frustrated that he wouldn't *listen* to me and became very agitated myself. He can be so stubborn! He talks like a damn attorney, pounding away at me like I'm on the witness stand. Before long I became totally emotional with tears streaming down my face. Rick tried to hold me but I pushed him away. After I calmed down a little we were able to talk things out.
>
> When we went to bed Rick wanted to have sex but I said "forget it" because I was still mad at him and didn't feel sexy at all. But in the morning I felt different. I opened my eyes and there was Rick's beautiful face, our noses almost touching. He gave me one of those big grins that made me fall in love with him and we embraced passionately. His penis felt wonderful as it pressed against my clit. What followed was surely one of our most amazing lovemaking sessions. I was *totally* excited. I couldn't get enough of him. I remained in a steamy mood all day long and we had boisterous sex again that night. Maybe we should fight more often.

Rick and Sheila each exhibit their individual anger styles. But because neither one attacks the integrity of the other—resisting any urge to go for the jugular—their basic trust and respect remain undiminished. Like many males, Rick (at least in Sheila's eyes) is relatively quick to anger, seems less reluctant to go on the offensive, and shows little inclination to back down (until Sheila cries). Men tend to feel anger more easily than they do other strong emotions even when they are obviously more hurt, frightened, or frustrated than angry. Anger is more compatible with the demands of the male role: to deny or conquer fear and to take an aggressive, competitive stance in the world.

Beginning early in life, most males are encouraged to develop antifear and antipain strategies, among the most potent of which is anger.

Sheila, on the other hand, is reluctant to express anger, and when she does, a torrent of other feelings flood her awareness and soon she's in tears—much to the consternation of Rick, who, like most men in his situation, has no idea what to do and probably feels manipulated, guilty, and robbed of his right to be angry. You can notice another common gender difference here as well: Rick is erotically energized by the fight and wants to make up by making love. He is aroused by the general excitation produced by his anger. Sheila, however, is turned off by anger and can't feel sexy until she calms down and reconnects with her loving feelings toward Rick the next morning.

By accentuating their differences, Rick and Sheila's fight helps to restore their individuality and separateness, opening up a desire-enhancing gap between them. As they pull back from each other to stake out their positions, they regain the perspective necessary to see afresh what drew them to each other in the first place. Their disagreement reestablishes the otherness that is the basis of all attraction.

ANGER AND THE FOUR CORNERSTONES

Anger's roles in erotic life are often subtle and easily missed. Consider again the cornerstones of eroticism, only this time pay special attention to the undercurrents of anger—some more visible than others—that often help to energize them.

Think back to a time when you remember longing for someone who wasn't easily available. When you finally did get together, I would guess that you felt joyful, not to mention aroused, and probably not at all angry—at least not consciously. If, however, your separations were extended and frequent, perhaps you resented the same unavailability that excited you. I've often wondered how much of the white-hot excitement of reuniting lovers is fueled by subtle resentment, especially when one partner feels more emotionally involved in the relationship and therefore longs more intensely. Tom, age twenty-two, describes such a situation. But unlike most longers, Tom is aware of his resentment:

> A couple years ago I dated an incredible fox who was fifteen years older than me. Not only was she beautiful, she also owned her own business and knew all the right moves in the sack. But it really bothered me that she was either traveling or in meetings just about every time I wanted to see her. I felt she was blowing me off even though she said she thought about me all the time.
>
> One night, as usual, I got her fucking answering machine and became totally pissed off. I drove over to her house and waited in her driveway until she finally pulled in about two hours later. She was glad to see me and we headed straight for the bedroom where I fucked her without mercy until she begged me to stop. She called me more often after that.

It's easy to see the dual meaning of "fuck" in Tom's story. His tolerance for longing, probably pretty low to begin with, is reduced even further by his insecurity about the relationship. Thus, in a style not at all uncommon for men, Tom converts his vulnerability into angry determination, which seems to have an aphrodisiac effect on both of them. One of the most important erotic uses of anger, especially for males, is the neutralization of anxiety. It's also easy to see the footprints of jealousy all over this story, another unpleasant but potent sexual intensifier in which anger is combined with insecurity.

You've seen how one of the cornerstones, searching for power, can intersect with arousal whenever you take control directly through sexual dominance or follow an indirect route to empowerment by submitting to an aggressive other. Although anger and aggression are often confused, aggression need not be related to anger, just as anger does not necessarily lead to aggression. Sometimes, however, the two do coincide so that the excitement of a top-bottom scenario is tinged with hostility. We'll take a closer look at hostility as an aphrodisiac when we explore the dark side of eros in the next chapter.

Both sexes often experience at least a hint of anger when they violate prohibitions. It asserts the primacy of the self over restrictive rules and regulations—and their perceived sources. More often than not, embedded in the joy of violational sex is an implied or direct "Fuck you" to prudes and naysayers everywhere. Anger says, "Your stupid rules don't apply to me; I'm in control of my sex life!" Thus it is a key emotion behind sexual acts or fantasies of defiance and rebellion.

THE RHYTHMS OF EMOTION

As you've focused on each of the emotional aphrodisiacs, you've probably been giving considerable thought to the feelings that have played a part in your own peak turn-ons. To facilitate your observations about the emotional aspects of your eroticism, I suggest that you use a simple rating scale from the SES to help give you a clearer picture of which emotions have energized each of your memorable encounters and fantasies. The scale ranges from zero (for emotions that were not at all important) to four (for emotions that were very important).

Begin by rating the importance of the response emotions in each of your peak turn-ons. As I mentioned at the beginning of the chapter, one or more of these feelings is mentioned by almost everyone in The Group. They are the emotional rewards of memorable sex:

Response Emotions

Not at all important [-0-1-2-3-4-] Very important

Exuberance

Related emotions: joy, celebration, surprise, freedom, euphoria, and pride.

Satisfaction

Related emotions: contentment, happiness, relaxation, and security.

Now rate the importance of the emotional aphrodisiacs in each of your most memorable encounters and fantasies:

Emotional Aphrodisiacs

Not at all important [0-1-2-3-4-] Very important

Closeness

Related emotions: love, tenderness, affection, connection, unity (oneness), and appreciation.

Anxiety

Related emotions: fear, vulnerability, weakness, worry, and nervousness.

Guilt

Related emotions: remorse, naughtiness, dirtiness, and shame.

Anger

Related emotions: hostility, contempt, hatred, resentment, and revenge.

These are the paradoxical emotions; they can either intensify or inhibit arousal. Consequently, rating them may be a bit more complicated. For one thing, the importance of any of the paradoxical emotions in peak sex is not necessarily related to its strength. This is especially true of the unexpected aphrodisiacs—anxiety, guilt, and anger—which are most likely to have had an aphrodisiac effect in their more subdued forms. Whereas few people have ever experienced "too much" satisfaction during or after peak sex, most of us have experienced too much anxiety, guilt, or anger. You may even recall instances when you became so emotionally enmeshed with a lover that your closeness turned you off.

FLUKES AND PREFERENCES

By rating the importance of each type of feeling in a variety of your peak turn-ons you will be able to determine whether certain emotions intensify your arousal consistently, only occasionally, or not at all. If your emotional ratings differ for each turn-on, this is an indication that your feelings were situational—simply the result of a unique set of circumstances. Remember Ron, who had a passionate and superanxious encounter in the basement of a London pub with his boss's wife? In this unique situation, anxiety whipped up his excitation to a fever pitch, whereas in other far less risky situations he felt too nervous to maintain an erection. Sometimes it's merely a fluke that a particular emotion makes a surprising appearance in an erotic scenario.

If many of your peak turn-ons have a high rating for one or more emotion, you can assume that you have a preference for that set of feelings; you consistently rely on them to help arouse you. Keep in mind that an emotion need not be felt strongly for it to be important to you erotically. For example, you might prefer to feel just a touch of guilt. Maybe if you're not feeling slightly naughty you become bored sexually. At the other extreme, if you are often flooded with intense guilt during or after sex, you are unlikely to experience it as an aphrodisiac.

If you're like most members of The Group, the unexpected aphrodisiacs—anxiety, guilt, and anger—are most likely to enhance arousal in low to moderate doses. Many members of The Group say that these "negative" aphrodisiacs appear in their peak turn-ons as often as the loving, celebrative, "positive" ones—but more in the background.

The closeness emotions are different and special. A sense of connection appears to be a predictable result of a high proportion of memorable turn-ons—even those that involve anxious or angry feelings. How can such negative emotions lead to such positive results? To understand this apparent contradiction, we must turn our attention to an extremely important characteristic of emotions—their changeability.

EMOTIONAL TRANSFORMATIONS

When you recognize and accept what you feel, without judging your emotions by logical standards, you will notice that the natural life of most feelings is remarkably short and fluid. If you are able to feel anger when you are threatened or when someone treats you unjustly, and if circumstances allow you to express yourself assertively, your anger will yield to a calm self-assurance. Likewise, if you're not ashamed to feel anxious when you perceive danger, chances are that you will take whatever steps are necessary to protect yourself, thereby demonstrating just how courageous you can be. Feelings that are attended to and honored move along, sometimes veering off in unexpected directions. It's the feelings that fester and won't let go that cause us distress. People who ignore or resist their feelings often end up obsessed with them.

The same principle holds true for emotional aphrodisiacs. They come and go, changing from one to another. In pleasurable erotic experiences feelings flow primarily from negative to positive. For example, when moderate anxiety (negative) adds intensity to a sexual scene, it typically results in a sense of security (positive). When, however, anxiety keeps building without being transformed into some other feeling, its unpleasant or disruptive aspects become increasingly pronounced and eventually undermine your enjoyment. The erotic significance of negative emotions lies mostly in their ability to shift at just the right moment. Whenever this change occurs—and it is usually not through conscious choice—a burst of erotic energy is released.

In her book about men's sexual fantasies, *Men in Love: The Triumph of Love over Rage,* Nancy Friday focuses on the tension and transformation brought about by conflicting emotions. Men desire approval and closeness from women, but counterbalancing this need is an accumulation of unconscious rage against women, whom Friday calls "the great sexual naysayers." She describes how men use sexual fantasy to transform frustration into love and acceptance:

> No matter what men do to/with their imaginary lovers, her reactions are just the opposite of mother's—*she loves him for it.* "Yes!" she shouts, "more!" A fantasy woman does not reproach her man for letting other men peep at her, for wanting to share her with another guy, for dreaming of her having sex with a dildo or a dog. Fantasy gives men the love of women they want, with none of the inhibiting feminine rules they hate. No matter how wild the man's sexual frenzy, the woman does not punish, but rewards. Love conquers rage.[15]

A similar reversal seems to lie at the heart of many women's fantasies. In a typical scenario an aggressive sexual male pushes through the heroine's resistance, releasing a wild, lusty woman from the confines of her normal reserve. As her full erotic powers blossom, she not only ignites a white-hot desire in her lover, but she also manages to transform his aggression into the perfect blend of animalism, sensitivity, and gentleness. Her pleasure becomes his primary concern. She shows him the path to love.

The transformation of one emotion into another, sometimes opposite one, is frequently more noticeable in fantasy than in real encounters. People who develop a comfortable acceptance of their erotic fantasies allow themselves to feel a wider range of emotions toward their fantasy partners than they do toward actual people. In addition, the unexpected aphrodisiacs are typically more easily recognized in fantasy because the fantasy itself is a safety factor. When you feel scared or angry in a fantasy, you imagine or remember feeling that way more than you actually feel it. Consequently, you can, for example, throw yourself into a dangerous or degrading fantasy scene you would never consider in real life because you know that you and your fantasy partner are both protected from harm.

Even though emotional transformations find their most dramatic expression in fantasy, peak encounters are also common settings for amazing emotional switch-overs. Once you recognize the changeability and fluidity of feelings, you will begin to notice that emotional transformations permeate erotic life. Here are the most common ways in which emotions redirect themselves during fulfilling sex:

Emotional Transformations

Anxiety ➨ Security

Weakness ➨ Strength

Guilt ➨ Freedom

Anger ➨ Appreciation

Emotional transformations play an important role in most of The Group's stories. Remember how Glynis overcame her persistent fear about having sex with another woman and found a sense of celebration? Or how Denise transformed the guilt she felt about having sex with her boss into defiant liberation?

Emotional aphrodisiacs follow their own rules, the first of which is that they refuse to be restricted by rationality. Whether for good or ill, feelings exist to be felt. To resist or deny our emotions is to strengthen them. True, the unpredictability of the emotional aphrodisiacs makes them potentially dangerous and bewildering. Trust your erotic mind, however, and even your least loving feelings may pull you circuitously toward pleasure and connection.

5

YOUR CORE EROTIC THEME

An internal blueprint for arousal transforms old wounds and conflicts into excitation.

When you contemplate the assortment of images and encounters that have aroused you, what do you see? There are those who perceive merely a random collection of events, each the result of a unique set of circumstances, separate and unrelated. Considering the effort you've invested in exploring your peak turn-ons, I suspect you've glimpsed recurring patterns among varied erotic experiences. But like most people, you're probably confused about what these patterns mean—or even what to call them.

Borrowing a term from the dramatic arts that everybody understands, sociologists William Simon and John Gagnon have proposed that we name these patterns "sexual scripts." Furthermore, they insist that our sexual fantasies and activities are influenced by these scripts to a far greater degree than most of us realize. They divide them into three basic types: (1) cultural scripts, (2) interpersonal scripts, and (3) intrapsychic (within the mind) scripts.[1]

All of us absorb an array of customs and traditions from our cultures, many having to do with sexuality. Invariably included are strong expectations for each gender along with deep-seated ideas about when, where, with whom, how, and how much sex is appropriate. Because cultural scripts are pervasive and begin impinging upon us from our first breaths, they become as much a part of who we are as our native language. Accordingly, they function automatically and are rarely questioned. Even sexual rebels are products of their cultures. They may violate society's norms and ideals, but they can only stake out their positions in relation to the very standards they're rejecting.

Whereas cultural scripts are totally disinterested in our individual preferences, interpersonal scripts are very different. Because they develop gradually as we learn the rules of social engagement, first within our families, and later with peers, our personalities play major roles in this learning process. As a result, even in the most restrictive cultures a variety of erotic styles coexist. Even so, members of subgroups within a culture often share a remarkable commonality in how they think about sex, who they consider attractive, and what roles they are drawn to play in sexual interactions.

Intrapsychic sexual scripts are the most idiosyncratic of all. Although inevitably influenced by cultural mores and interpersonal norms, intrapsychic scripts are, first and foremost, expressions of each person's unique responses to his or her life experience—beginning as a small child. As a result, there are an infinite number of intrapsychic scripts in the world. I'm particularly fascinated by intrapsychic scripts because they reflect the eccentricities of each individual's erotic mind more so than any other kind. But the sheer volume of these scripts makes it extremely difficult to identify which, if any, common threads connect them. This is why I've borrowed another term from the world of drama: themes.

An easy way to distinguish between scripts and themes is to recall the enormously popular television series "Columbo," still enjoyed in reruns by millions around the world. The theme of the show is always the same: a murder is committed on camera, so we, the viewers, know who did it, how it was done, and the motive for it. We watch the rest of the show not to find out whodunit, but to see how the disheveled lieutenant will identify and corner the killer. This same theme is repeated in every episode, yet the script is always different—new characters, motives, methods, and slip-ups.

Themes are the underlying forms from which scripts unfold. Whereas scripts tend to be detailed, themes are simple and can often be described in a sentence or two, sometimes a single phrase. My studies have led me to conclude that the existence of themes is what gives us that sense of déjà vu in our erotic lives. Most of us act out many sexual scripts, but only a few themes.

In this chapter I am going even further by proposing that our most compelling turn-ons are shaped by one unifying scenario that I call the core erotic theme (CET).[2] The CET is core because it occupies a place at the heart of each individual's eroticism. And it's thematic in the sense that an infinite array of storylines, characters, and plot twists can all be inspired by a simple, yet profoundly meaningful, dramatic concept. If you wish to touch the deepest sources of your eroticism, delve into your CET, for it is the most ingenious invention of your erotic mind.

Your CET begins its long evolution during childhood and is first sketched out in fantasies and daydreams you probably don't remember. Because these early images almost certainly grew out of impulses and interests considered inappropriate for children, they were veiled in secrecy. Even now you probably still keep certain ultra-personal turn-ons—those that spring from your CET—hidden from other people and quite possibly even from yourself.[3] To whatever extent you feel comfortable, take the risk of exploring your CET. Its significance is so vast that even small discoveries about it can be highly revealing and useful.

At the most fundamental level, your CET is an amazingly efficient shorthand encapsulating crucial lessons about which people, situations, and images tend to evoke your most forceful genital and psychic responses. The CET, however, is far more than a mere checklist of what and who turns you on. Its extraordinary power arises from the fact that it links today's compelling turn-ons with crucial challenges and difficulties from your past. Hidden within your CET is a formula for transforming unfinished emotional business from childhood and adolescence into excitation and pleasure.

The same peak turn-ons that have already yielded so much information about the inner workings of your eroticism are also rich with clues about your CET. As you ponder an exciting experience, looking beyond the captivating details and thrilling sensations, try to see why these experiences were so exciting. Look closely enough and you'll undoubtedly find subtle reminders of one or more of your most vexatious problems. Although it may seem illogical that exciting sex should have anything to do with life's unresolved struggles, one of the most important insights you can have about the erotic mind is that *high states of arousal flow from the tension between persistent problems and triumphant solutions.*

You can enjoy sex without giving any thought to your CET. In most cases the scripts and themes that guide erotic life perform their functions subconsciously. In fact, some people have told me in no uncertain terms that they prefer not to know about the deeper meanings of their hottest turn-ons. I've noticed, however, that those who study their CETs consistently develop a new level of respect for their eroticism and a greater ability to understand and influence their sexual choices. This chapter is designed so you can choose the level of awareness that feels most comfortable. You may read it either as an examination of other people's sexual quirks and eccentricities or as an opportunity to look more closely at your own. I suggest you do both.

SEXUAL HEALING

Even though your eroticism subtly reflects the challenges you faced while growing up, when you're caught up in the thrill of escalating arousal and orgasm you aren't consciously thinking about these problems; your attention is riveted on the pleasures of the moment. The fact that you are excited shows that your CET is working. After all, the purpose of your CET is to use old wounds and conflicts as aphrodisiacs.

The relationship between someone's CET and his or her difficulties from the past isn't easy to recognize—a fact I've repeatedly encountered as a therapist. Similarly, in spite of The Group's incredible openness in describing their peak turn-ons, when the Sexual Excitement Survey asks for their ideas about what made these experiences so arousing, most summarize what they've already said, without venturing into deeper territory. But some unusually perceptive members of The Group spontaneously offer insightful comments about connections they see between their peak experiences and significant lifelong dilemmas.

Jana: Object of pursuit

Jana, a thirty-seven-year-old public health nurse, describes her fascination with the process of selecting memorable turn-ons. Her curiosity uncovers a common thread of which she had previously been unaware:

> I had trouble picking just two exciting encounters so I made a list of them all. It dawned on me that in virtually every case I'm being aggressively pursued by a handsome and determined man. My role is to act rather coy and passive, as if I want them to *prove* their interest in me through sheer persistence.
>
> I had never seen this so clearly before because I'm usually obsessed with how handsome the man is or how big or strong. I've never stopped to question what *I'm* feeling. But once I saw my taste for being pursued I couldn't stop thinking about it and even brought the subject up with my therapist.

Most people begin exploring their CETs much as Jana did. First they'll notice a recurring motif—in Jana's case, her need to be pursued. Next they become curious about its broader meaning. They wonder, "How has this theme played itself out in my life?" Jana recognizes that similar scenarios interested her as a girl:

> I remember the desire to be pursued in sexual fantasies as far back as age eight or nine, maybe before. I use feminine poses to attract a rich and famous man. But because I'm so shy and reserved he's "forced" to seduce me. Once I surrender he whisks me away on his yacht or horse and I feel chosen and very special. In all my fantasies today, and my best encounters too, I feel *exactly* the same way.
>
> The imperative of feeling desirable stands out for me because in reality I've *never* seen myself as attractive. On the contrary, I've always wished I were as pretty and sophisti-

cated as my older sister. She got all the attention from guys, teachers—everyone. I was an awkward "tomboy" and I believed my parents liked her better. I remember crying myself to sleep over my fate. Now I know intellectually that I'm not ugly, but I still think of myself that way. I'm always trying to fix this by getting men to want me.

If I surrender too quickly it's not nearly so exciting as when I get the full seduction treatment. It makes me feel feminine and beguiling to be chased. I imagine they can't resist me. Now that I live with a wonderful man I'm still always waiting for him to initiate sex (which he complains about a lot). It's hard to admit, but when he comes on to me forcefully it's almost like getting even with my sister who I both loved and hated for being so damned perfect. When it all works—which it always does in fantasy and occasionally in reality—I'm getting the attention I've craved all my life.

As is usually the case, the goal of Jana's CET is simple and straightforward: to demonstrate her worth by being the object of pursuit. Her CET is a formula for soothing the hurt of perceiving herself as ugly, inferior, second-best. At the same time, her CET reverses the outcome of the painful rivalry she felt with her sister because she wins; now *she's* the desired one.

Within its simple framework, a CET brings together all the varied aspects of erotic life we have explored thus far. Notice, for instance, how Jana makes her pursuers surmount obstacles to get to her. Without consciously realizing it she's been using the erotic equation. In addition, virtually every CET I've heard of rests on one or more of the four cornerstones of eroticism. For Jana, the drama of searching for power is played out when she demonstrates her control by "forcing" men to pursue her. By winning the prize of an appealing man, she temporarily conquers her feelings of inferiority.

I've consistently noticed that in pursuer-pursued scenarios—among the most common of all CETs—overcoming ambivalence also plays a central role. Although Jana doesn't say so, I would guess that she experiences ambivalence toward her pursuers similarly to what she always has felt toward her sister. After all, her worth is on the line; she's at their mercy. In addition, just below the surface of Jana's story flows a deep undercurrent of longing for the love and respect that came so easily to her sister.

Every CET is also energized by one or more emotional aphrodisiacs. Jana is courageously honest in recognizing that she enjoys a sense of revenge against her sister as her own value is acclaimed. I would also speculate that when it's unclear if the pursuer will be sufficiently persistent, Jana's anxiety level escalates considerably.

SEARCHING FOR WHOLENESS

Not only is the CET a creative strategy for transforming emotional pain into excitation, it's also an expression of a quest for wholeness and completion. CETs help us select partners who value our strengths and compensate for our weaknesses, crucial aspects of the "chemistry" of attraction. Through a process Dr. Tripp calls "exporting," our self-esteem is boosted when someone wants what we have to offer. It can be particularly rewarding to be desired for qualities about which we normally feel insecure.

Through a complementary process called "importing," we zero in on characteristics of the other that fill in our own missing or underdeveloped aspects. Whereas the goal of exporting is personal validation, the purpose of importing is the pursuit of wholeness. When we feel attracted, few of us are consciously thinking, "This person is desirable because he or she exudes the qualities I lack." But to understand as much as possible about our attractions, we must look at how we feel deficient or out of balance—not always a pleasant task, but an extremely important one.

Claude: Awakening the lioness

It's easy to see why Claude, an economist in his mid-thirties, feels so good about this peak encounter:

> I attended a party where I knew only the host. While sitting in a corner I spotted a new arrival, a smartly dressed, petite woman with jet black hair and delicate features. I decided to venture out and introduce myself. She was soft-spoken but

> friendly and highly intelligent about many topics. We talked until the party was breaking up. Finally, I mustered the courage to invite her home for coffee and she accepted.
>
> After more conversation we began kissing. Her lips were soft and sensuous. Soon we were fondling each other quite freely, still fully clothed. I laid out a comforter in front of the fireplace where we stripped each other as our kissing and groping grew more urgent. Much to my surprise and delight this soft-spoken lady became a lioness in heat. When I sucked her nipples she groaned and writhed. And when I buried my face in her bush she sighed and whimpered and screamed.
>
> Obviously, she was losing control which turned me on immensely. This little doll became overwhelmed with passion. I couldn't believe I lasted so long because her screams excited me so. It was unbelievable when she came. Her spasms went on and on, rippling through her body like earthquakes. She was the most passionate, loudest, and uninhibited comer I'd ever been with. We repeated an equally noisy romp the next morning. Since then we've talked a couple of times on the phone but we never met again. I know I won't forget her, though.

Judging from this story, Claude's CET involves being the catalyst for escalating passion in his partner. Millions of men—and more than a few women—can empathize with Claude's delight at releasing a wild, sex-crazed animal from a reserved, soft-spoken lady. He is exporting masculine erotic vigor. And the fact that his partner responds so enthusiastically confirms his virility:

> She might have reacted the same to every guy, but to me her wildness was my personal cheering section. Her screams and moans made me feel like a terrific lover. In fact, all of my best sex has been with women like this—hidden lionesses. My super hard-on and staying power showed how confident she made me.

So common is this theme among men that we could easily decide that no further analysis is necessary. But Claude reveals another dimension to his arousal, one not visible on the surface:

> Both in and out of bed I've been called the "strong, silent type." Women always want me to say more about what I feel, to be more passionate and less cool. Even when I come I hardly make a sound. I've always picked women who have emotions about *everything*. I'm sure my attraction has to do with my difficulty expressing myself. It's almost as if they feel the passion for *both* of us. I admire these women and wish I could be more like them.

By stimulating a lover to be demonstrative in ways he can't, he maintains a profoundly exciting contrast with his lovers and exposes himself to characteristics he lacks. As it is with Claude, attractive others are, at least in part, mirrors in whom we perceive underdeveloped or missing aspects of our own personalities. Mature lovers recognize that intimate involvements are opportunities for growing. Through our attachments we can gradually cultivate within ourselves the very characteristics we find so appealing in our partners.

THE ROYAL ROAD TO YOUR CET

Jana and Claude's insights demonstrate how much can be revealed by patiently probing your real-life attractions and encounters. There are, however, limitations to this method. It is important to keep in mind that your CET exists and operates internally even when it is expressed externally through your choices of partners and preferred sexual practices. During partner sex your emphasis is on the interplay between you. Unless both of you are unusually forthright in acting out your innermost desires, the niceties of social-sexual interaction can easily divert you from the unedited content of your CET.

Awareness of your CET during partner sex is also limited by the fact that it probably includes one or more aspects you don't want to act out with a real partner. Quite naturally, you conceal these extremely personal aspects of your CET. In my experience the CET can be most freely explored when it need not be negotiated to mesh with the needs of another. Simon and Gagnon rightly insist that "the sexual dialogue with the other often bears little resemblance to the sexual dialogue with the self."[4]

Your CET, with its intimate connection to your deepest and often hidden concerns, has much in common with dreams. Emanating from the subconscious world beyond the constraints of logic, social obligations, and morality, the dream is a canvas upon which anything can be painted. Just as Freud declared the unfettered imagery of dreams to be a "royal road to a knowledge of the unconscious activities or the mind,"[5] sexual fantasies and daydreams are the royal road to your CET. When you allow your fantasies free rein, especially the ones you repeatedly gravitate toward during masturbation, "pure" representations of your CET hover closer to consciousness than at any other time.

Perhaps this explains two fascinating findings from the SES. After respondents write about their favorite fantasies, I ask them to estimate the proportion of other arousing fantasies they think are based on similar themes. Almost two-thirds of The Group say that at least 80 percent of all their fantasies closely resemble their favorite one! Equally striking is the fact that almost as large a proportion recall having similar fantasies for ten or more years, often for as long as they can remember. In other words, although fantasy scenarios vary tremendously from person to person, and even within the same person, there is a high degree of repetition when it comes to our most compelling ones—which argues for the influence of an underlying motif.

UNRECOGNIZED FANTASIES

Those who give themselves permission to examine their fantasies usually find it easier to identify the themes from which those images derive their erotic power. But what if you fantasize very little or not at all? You can still explore, although in subtler form, the dramatic elements that animate your passions by paying close attention to the content of your strongest attractions as well as the features of real-life encounters that produce high levels of arousal.

But before you abandon fantasy as a potentially rich source of information about your CET, I recommend that you reconsider your beliefs about fantasy. In most cases I've found that those who think they rarely or never fantasize are similar to people who don't remember dreaming. They're simply not in the habit of paying sufficiently close attention. Consider a discussion I had with Lorna, who came to me for therapy because she usually lubricated very little during intercourse with her husband, Mike, resulting in irritation rather than pleasure. She realized her lack of lubrication reflected a lack of arousal, so I asked if she ever used fantasy to help turn herself on:

Lorna: I don't really have sexual fantasies.

Jack: You mentioned you were necking with Mike last week and you felt aroused. What was going through your mind then?

Lorna: Nothing in particular. Just feeling nice—warm and tingly.

Jack: Did you notice any thoughts or feelings along with those wonderful sensations?

Lorna: (after a long silence) Well, I vaguely remember noticing how muscular and hard his back was as I held on. I've always liked his muscles.

Jack: What about them?

Lorna: Oh, you know, I like remembering how masculine he was when we first met. He was hairy and big and I soft and naive. I've always liked the contrast.

Jack: Did that memory help turn you on?

Lorna: I suppose it did. The differences are still there, even though I'm not as naive anymore.

Jack: Lorna, you may not think so, but what you just described to me is a fantasy. It all took place in your mind.

Sexy thoughts that aren't considered fantasy are especially common among women. In Lorna's case, recognizing that she *had* fantasies allowed her to focus on the images that turned her on, which in turn gradually increased her arousal and thus her lubrication. As with many people, Lorna's fantasies were mere snippets of erotic thought, although these fragments grew richer as she acknowledged their existence. Eventually, she discovered that if she imagined herself, innocent and naive, being taken by a strong and virile Mike, her arousal would intensify—quite impressive for a woman who didn't fantasize!

The amazing capacity of the CET to encapsulate an incredible amount of detail is one reason that Lorna didn't readily perceive herself as a fantasizer. Her experience demonstrates the importance of *erotic cues* in our CETs. These cues are extremely subtle and specific, perhaps a particular gesture, a stance, the shape of a single body part, or a certain look. Whenever Lorna focused on Mike's muscles and hairy body, these characteristics served as a supercondensed shorthand for her fascination with the contrast between masculinity and femininity.

EXPLORING YOUR CET WITH FANTASY

Men and women with active fantasy lives are usually quite familiar with the sorts of partners and situations likely to electrify them. In most cases exceptionally exciting fantasies are closely related to a person's CET. The challenge is to discover the links between these fantasies and the unfinished emotional business they're trying to resolve.

Although we'll focus on just a handful of specific themes in this chapter, it is not my intention to limit what you might discover for yourself. After all, the possibilities are practically limitless.[6] I've chosen examples that I hope will both pique your interest and demonstrate how the erotic mind is involved in all spheres of life.

To illustrate how fantasy can illuminate a CET's deeper significance, I want to highlight a theme that I had an opportunity to explore with a client in therapy. An ongoing therapeutic dialogue is often the best avenue for peeling back layers of meaning. This scenario revolves around shocking or overwhelming an unsuspecting partner with a voracious passion that the fantasizer normally keeps hidden. Judging from The Group's favorite fantasies, as well as those I hear from many other clients, it's quite a popular concept. But what gives this theme its power to arouse?

Felicia: The nympho within

Considering that CETs begin forming early in our lives, it should come as no surprise that many scenarios revolve around having a secret "alter-ego," a part of oneself with the freedom to be unabashedly erotic in ways that were impossible as a child. Felicia, a forty-one-year-old social worker, reveals hers:

> I invite a proper gentleman over for tea. He's seen me only in my role as a dedicated professional and assumes this will be just a friendly date. I maintain an air of formality until I can't wait to let the nymphomaniac in me come out. I casually unbutton my blouse (wearing no bra, of course). He's obviously taken aback, especially when I move in closer.
>
> "Why don't I loosen that tie?" I ask, not waiting for a response before I remove it and slowly unbutton his shirt. I can feel his nervous excitement. I gently rub his crotch. "Felicia, I . . ." he protests, but I cover his mouth as I unzip and go down on him. I'm in total control. He's under the spell of my expert lips and tongue. When I feel him getting extra hard just before he's about to come, I stop until he calms down a little.
>
> Then I really get loose, brazenly undressing both of us. I know I'm blowing him away because he *never* expected this from me. I lay him out on the sofa and fuck him wildly with me on top. I'm holding his arms down, kissing him and rubbing my breasts all over him. Soon he's thrusting as wildly as I am until I feel him explode inside me.
>
> I keep him inside even after he starts to go soft. I lay still except for licking his nipples. As he starts to rise again I begin moving ever so slightly, milking his dick with my expert vagina. I release his hands so he can rub my clit as we fuck again.
>
> After dressing in silence we both assume our proper roles and I escort him to the door. He doesn't know what's hit him but I'm sure he'll be back for more.

Felicia's delight in shocking her unsuspecting partner as she reveals the raw energy of her sexuality is common among both sexes, but especially women. She goes on to explain the special significance of letting out what she calls "the nympho within":

> This fantasy has always given me such a thrill because it's a secret me that hardly anyone sees. I take after my mom who's always worried what others think. She's so uptight. But I was a sexy girl, always very interested in naked bodies. I knew by her complete silence on the subject that Mom *hated* sex. I had no choice but to adopt her demeanor.
>
> On the surface I gave in to her by becoming the perfect child. But I wasn't about to let her de-sex me! In my fantasies I'm as wild as I want to be. Unfortunately, I wish I could be more that way with my husband. I like sex with him but I'm rather inhibited much of the time. But if it weren't for my fantasies I'd probably be as asexual as my mom.

Felicia understands the role of her CET: to act out without reservation the eroticism that she suppressed in deference to her mother. Her CET both expresses and solves her predicament as a sexy girl growing up in an antisexual environment. In her secret fantasy life she nurtures erotic vitality while providing an outlet for the repressed aspects of her personality.

PORNOGRAPHY, EROTICA, AND YOUR CET

We've seen how the CET, as a product of the imagination, is often expressed most clearly in fantasy. It's not unusual, though, for both men and women to use external stimulation such as sexually explicit stories or pictures to ignite their imaginations. If you enjoy these materials, studying the ones that move you may give you valuable insights into the content of your CET.

Visual porn, the most popular kind produced by and for men, is relatively generic in the sense that a succession of sexy acts can serve as erotic cues for a wide range of fantasies. The focus is on raw, unencumbered lust. Toward that end, most male porn makes a point of creating a sleazy atmosphere to set it apart clearly from everyday reality. Also common are variations on themes of dominance and submission, male prowess (symbolized by huge genitals and buckets of semen), and group sex with two or more women, often including sexual interactions between the women—all for the entertainment of the man, of course. But the primary focus is on erogenous zones in states of feverish interaction.

Pornography produced by and for women, while not devoid of wanton lust, sleaze, or power scenes, virtually always has a *context.* There is a lead-in to generate a mood and anchor the characters in at least a minimal relationship, even if coercion or rape will become part of the story. Although women may want their erotica to have a plot, often a romantic one, research has shown that when it comes to producing genital arousal, explicit sex is what turns women on—just as it does men. But women don't always realize they're aroused.[7]

Erotic materials can only shed light on your CET if they help you pinpoint exactly what arouses you. If the erotica is visual, notice the body types of the characters who fascinate you and contemplate what these bodies represent. What attitudes do the characters exude? What is it about these attitudes that stimulates you? How do the characters interact? What do you imagine they are thinking and feeling? This process works best if you can relate what you see to your own fantasies. For some people, especially men who consistently masturbate to visual porn, explicit images can actually distract them from their CETs. They feel highly turned on but without any idea why.[8]

Written stories contain more detail and therefore offer greater opportunities for studying specific characters and scenes that coincide with your CET. If the idea appeals to you, select one or more of the many collections of sexy stories or fantasies written by men, women, or both.[9] As you read them, notice which ones generate a strong erotic response, without worrying about why. Keep in mind, however, that you are likely to feel arousal in response to many scenes, some of which may surprise you. Just because you're turned on doesn't necessarily mean your CET is involved. But once you identify several stories that excite you consistently and powerfully, make a list of features they have in common. This list will contain invaluable clues about your CET.

SHADOW THEMES

The long process of learning which feelings, thoughts, and behaviors are acceptable and which are not begins in infancy. During our psychological development we repudiate certain aspects of ourselves that we fear might lead to disapproval or abandonment by those we depend upon for survival. Gradually, images and impulses that are denied conscious recognition coalesce into a portion of the unconscious that Carl Jung aptly named "the shadow." Although we may try to ignore the shadow, it is a universal dimension of life and therefore of eros.

Having been banished from consciousness, shadow impulses tend to fester, assuming exaggerated and distorted proportions—which is why the shadow so often breaks into consciousness in the form of bizarre or "perverted" sexual fantasies that flagrantly disregard both cultural norms and personal values. Shadow fantasies commonly involve manipulation, exploitation, coercion, and a host of other violations, both large and small. Sexologists now realize that such fantasies are normal expressions of the erotic imagination.

Shadow fantasies, even vile ones, rarely cause harm. Most people don't really want to act out their darker fantasies, especially in ways that are clearly damaging. But many thoroughly enjoy consensual role playing in which all kinds of violations are reenacted, sometimes complete with elaborate props. Problems most commonly erupt in those who keep the shadow locked away in its unconscious dungeon, thus eliminating any chance for them to reintegrate their rejected aspects into an expanded self-awareness. Jung insisted that learning to accept one's shadow is absolutely essential for psychological wholeness:

> One does not become enlightened by imagining figures of light, but by making the darkness conscious. The later procedure, however, is disagreeable and therefore not popular.[10]

The shadow, including its erotic manifestations, holds the key to the whole self, as opposed to the limited self to which most of us have become accustomed. When shadow material is denied an outlet, pressure builds for it to push beyond the safe boundaries of the imagination and become destructive. The shadow is darkest when we refuse to look at it. In this section our goal is to gaze at the shadow of eros to comprehend its motives.

EROS AND THE HOSTILE IMAGINATION

In a remarkable book, *Sexual Excitement,* psychoanalyst Robert Stoller makes a stunning assertion:

> Putting aside the obvious effects that result from direct stimulation of erotic body parts, *it is hostility—the desire, overt or hidden, to harm another person—that generates and enhances sexual excitement* [italics added]. The absence of hostility leads to sexual indifference and boredom. The hostility of erotism [*sic*] is an attempt, repeated over and over, to undo childhood traumas and frustrations that threatened the development of one's masculinity and femininity.[11]

Stoller's emphasis on hostility as the motivating force behind high excitement is difficult to accept, although it is a bit easier for us because we have already become familiar with anger as an emotional aphrodisiac. I'm sure you can also see Stoller's contribution to my concept of the CET as a formula for transforming unfinished emotional business into arousal. However, I disagree with him on several points.

I believe hostility is just one of many emotions that energize our eroticism and not necessarily the most important. In addition, Stoller has little to say about how readily hostility can coexist with, or transform into, positive feelings. Finally, I believe that *any* threats to our self-esteem—not just those involving gender identity—can become intertwined with our eroticism. But I strongly concur that to understand our deepest erotic impulses, especially our strategies for undoing traumas from the past, we must face our psychic wounds, and with them our hidden desire to avenge ourselves.

Nadine: Turning the tables

Several members of The Group recognize the presence of hostility in their peak turn-ons. Nadine, a graduate student of psychology in her mid-thirties, confronts her shadow as she recounts this fantasy:

> I'm a pubescent girl in the care of a middle-aged man I don't know. He takes good care of me but forces me to have sex with him and his friends and strangers. They examine, fondle, and praise my body parts. They strip me while they're dressed except for their penises sticking out. They touch and prod me everywhere—my mouth, ass, pussy. Some squeeze my breasts while others are jacking off and coming all over me. Before long I'm dripping with semen.
>
> I have to be available at all times to be taken and used by this man or anyone else he pleases. These are always rough, disgusting men. I am naked and totally exposed. I vary my fantasies as much as possible. Sometimes we are on trains, in back alleys, or even in hidden caverns of a church. The men are interchangeable because I don't care about their faces. But I always feel forced and exposed. In this sense my fantasies are essentially the same as when I first imagined being kidnapped by pirates as a little girl.[11]

For at least twenty-five years Nadine's CET has drawn its energy from her being overpowered, exposed, and used by distasteful men. When she offers her ideas about the meaning of her fantasy, she first mentions a paradox we have encountered before: her tormentors demonstrate her desirability by being so turned on by her that they *must* have and use her; they can't help themselves. Her complete submission also places her at the center of attention. Nadine has additional insights about the key themes of helplessness and exposure:

> The submission idea has always held a great appeal. As a girl I obviously couldn't be responsible for such disgusting wishes so I always needed outside aggressors to make things happen. I suppose this is a prime reason why I prefer to be a budding young woman in my fantasies, supposedly innocent—but not really, of course.
>
> But more than being overpowered I relish the *exposure*, having every part of me examined in detail. In a strange way, this scrutiny reminds me of the hours I used to spend exam-

ining *myself* as a girl, usually feeling horrible about being overweight and obsessed with zits and other imperfections. Going to school was an ordeal because I couldn't hide my fat or face. I worried about teasing far more than it actually happened. In my fantasy I'm trim and gorgeous. Even though I haven't been fat since my teens, I still feel that way.

I see how many problems get handled through this scene, though I know it appears to be torture and abuse. I feel so hot and sexy and wonderfully helpless and everyone does what I want. My "caretaker" sets up everything for me and makes me feel safe, even though I don't know him because I would *never* bring anyone who knows me into a fantasy like this.

I'm not sure why I pick such gross men. I don't like macho guys, preferring more sensitive ones. These are the men I love but not the ones who excite me in fantasy. I think it has to do with how much I resent the way macho types see women only as sex objects. Maybe it's odd that I love being an object in fantasy. But I'm actually making *them* objects—I give them no real identity of their own. They are simply the tools of my pleasure—*dispensable nobodies.*

Through self-analysis Nadine has stumbled upon one of the most widespread shadow themes—and one of special interest to Dr. Stoller—dehumanization. He has described how the hostile imagination often robs its objects of their personhood. In a sense, Nadine's coarse men are fetish objects whom she can manipulate by having them manipulate her. In the process her pain is magically transformed into excitement.

But is she healed by her turn-on or simply repeating the same old story and paying a higher price for the privilege than she realizes? How can anyone who wants to be treated so badly possibly feel good about herself? These and other questions will occupy us in Part II, "Troublesome Turn-ons." One thing I know: many people *are* healed—at least temporarily—by such fantasies. They wouldn't be nearly so widespread if they didn't have much to offer.

THE PUZZLE OF SADOMASOCHISM

Two centuries ago, the infamous Marquis de Sade laid out his dark vision of sex and human nature in a deliberately odious collection of novels, plays, and essays that remain the quintessential literature of the sexual shadow. In his provocative book *Dark Eros*, Thomas Moore describes Sade's unsavory aesthetic:

> Sade seems to have fixed his vision on this underworld with the tenacity and intensity you might find in mystical literature. He approaches humankind's love of evil with the devotion and faithfulness of a saint, and so with good reason he has been called "The Divine Marquis."[12]
>
> Sade's hatred of convention and sentimentality is not just personal misanthropy. It is a positive recognition of the dangers of pleasantness taken to the extreme and made into a rule. Sade's leather straps reflect as in a mirror the ways in which the conventional world can handcuff a life.[13]

In the Sadian aesthetic, as is so often the case in the shadow world, the usual order of things is reversed. He revels in everything considered disgusting, perverse, or depraved: cruelty is a virtue, degradation is a sacred act, feces are objects of enthusiasm, and pain is pleasure.

Sade is best known for a form of kinky sex that bears his name—sadism, which is one side of a unitary phenomenon known as sadomasochism (S-M in popular parlance). Sadomasochists create a unique kind of erotic intensity by inflicting and/or receiving physical or emotional pain. The pain may range from very mild to extreme but it always symbolizes eroticized power, a combination with which we have become familiar. Most S-M impulses are expressed in fantasy, but many are also acted out in role playing—often with paraphernalia such as whips or any of a variety of devices for bondage, restraint, and punishment.

Some of the pain inflicted or enjoyed in wild sex is simply a response to escalating passions. Not only might you feel inclined to bite, scratch, pinch, or slap when highly aroused, but at the same time, you're less likely to be hurt by these activities because your pain threshold actually goes up in the heat of passion. In other words, things that would normally hurt can feel terrific when you're excited. You may discover extra intense sensation at the changeable boundary between pleasure and pain. This intensity doesn't necessarily have much to do with S-M. However, if you've ever enjoyed the point at which pleasure and pain begin to blur, you'll probably find it a bit easier to understand how pain can become a turn-on.

Like all CETs, S-M scenarios are primarily concerned with the resolution of childhood conflicts and hurt. But instead of addressing hurtful problems in disguised or subtle ways as most CETs do, sadomasochism turns the spotlight on pain and provides two complementary strategies for mastering it.

The sadist takes command of his or her psychic wounds by skillfully administering pain to an enthusiastic recipient. The sadist is spared the discomfort of the hurt and is also gratified and subconsciously relieved to observe that the masochist clearly likes it. Old wounds are simultaneously avenged and transformed into an erotic high. The sadist is beyond merely being safe and has the illusion of omnipotence.

For many people the allure of masochism is incomprehensible. Yet sexologists have long recognized that more people imagine or play the masochistic role than the sadistic role, a preference that is clearly evident in The Group's peak turn-ons. What could anyone possibly gain from being hurt or humiliated? And if resolving pain is a chief goal of our CETs, why would someone seek it out? Dr. Stoller explains a crucial piece of the puzzle:

> Masochism is a technique of control, first discovered in childhood following trauma, the onslaught of the unexpected. The child believes it can prevent further trauma by reenacting the original trauma. Then, as master of the script, he is no longer a victim; he can decide for himself when to suffer pain rather than having it strike without warning.[14]

The notion that masochistic scenarios help participants seize control of early trauma is supported by the fact that masochists frequently report feeling validated and powerful—just the opposite of what you might expect. Psychologists who have analyzed S-M dynamics have noted that the masochist is the true master, often aggressively choreographing the entire encounter.[15]

Interestingly, The Group doesn't report any peak experiences, not even fantasies, in which they enjoy inflicting physical pain on their partners. For reasonably healthy people, actually hurting someone is rarely, if ever, a peak experience. Symbolically hurting them, however, is a different matter altogether.

Web: Defiling innocence

The erotic imagination is fascinated by the contrast between innocent purity and wanton violations. Not only does the ravisher require an unspoiled object, but the unjaded freshness of naive, virginal characters seems to cry out to be defiled. And if you believe urges to trample on innocence are solely the invention of depraved adults, think again. When The Group recalls their earliest fantasies, many of the most common, especially among women, involve being corrupted by faceless strangers or virtuous authority figures, such as priests, nuns, teachers, or doctors.

The joy of defilement is at the center of Web's favorite fantasy. This sixty-one-year-old retired executive is aware of the all-important reciprocity between the defiler and the defiled:

> The fantasy that stirs me most is a scene where a young virgin confides in me about a problem with her overhorny boyfriend. I tell her I will only help her if she follows my directions without question. She agrees. I tie her against a pillar and slowly caress her while she both protests and groans.
>
> Then I shock her by ripping off her cute little dress, tossing it on the floor in a shredded heap. I like that she's afraid but I reassure her with my eyes and a slight smile. She surrenders completely and her body comes *alive* with pleasure. As her teacher I fuck her mercilessly—but also with love.

As is so often the case, at the dramatic moment when reticence is transformed into enthusiastic compliance, a burst of erotic energy is released. Web shows considerable insight into what he's doing and why:

> In a way I'm reliving my younger years and all the frustration I felt about being sexually restrained as a child and then later with sexually uptight girls (back then virginity was

popular). In my fantasy, I'm obviously manipulating and mistreating her. But I'm also worshipping the purity I must destroy so she can become the woman I crave.

His CET is animated by a complex concoction of hostility, revenge, and an abiding affection. The central dilemma is Web's love-hate relationship with innocence, which simultaneously attracts and frustrates him. Through years of experimentation he has perfected an imaginative formula in which this long-standing conflict generates high erotic intensity. He begins with the unspoiled purity that fascinates him—a slut, of course, would never do. But instead of feeling rejected by innocence as he has in the past, he corrupts it. And through the defilement, passion is released in both participants. Web's CET is touched by the shadow but is neither cruel nor malicious.

Margo: The joy of humiliation

The paradox of masochism, wherein one finds pleasure and empowerment by being put down or humiliated by a sadistic partner, is evident in a peak encounter described by Margo, an articulate, introspective sexual adventurer who has been earning her way through college as an erotic dancer:

> I often perform at bachelor parties and am excited all evening, like constant foreplay. I am very sensual and love it, although I understand the difference between sexual titillation and love and sex with my boyfriend. One evening I was uncontrollably turned on by a guy at a party (instead of my usual "tease" which has a certain distance to it).
>
> He was gorgeous, an arrogant, sexy, cold model—exactly the type I *dislike,* so egotistical and smug in his sexiness. But it was pure animal attraction and I felt a surge of energy. All night I came back to him and when I kissed him it was real and everyone in the place could see it. Later I loved it when he grabbed my hand and dragged me (no words spoken) into a storage room and watched—detached—as I went down on him. I was drawn to his icy ways, to his restrained aggression, and to his incredible physical beauty.

> After the party, he brought me home to his bed and denied me his cock, although I knew he wanted me. I felt extreme passion in the denial. He treated me coldly like the whore I wanted to be. He held my head down and spoke to me in the language of hot, hot lust. "I want your juicy cunt you nasty bitch," he said with disdain. It was egotistical sex, two people flattering and demeaning each other. I was slave *and* master.
>
> When I got home very late and told my boyfriend I had "slipped," he was so angry and completely disgusted he turned cold. It hurt to be treated that way but I was also incredibly excited when he ordered me to get down on my knees and said, "Did he like how you suck cock, too?" It was emotional S-M and much more exciting than actual physical bondage ever could be.
>
> I liked the freedom to experience the dark side of myself. To be a whore, just a body, fulfilling each other's nastiest fantasies. I wanted to be hurt, denied, insulted and he wanted control.

Judging by the intensity of her response, we can reasonably assume that Margo's CET energizes these encounters. But what unseen needs make Margo enjoy a double dose of humiliation and mistreatment? Is this the price she must pay for being a sex-crazed erotic dancer? If so, she shows barely a trace of suffering. Instead the hostility of her cold and arrogant customer, and later the retaliatory anger of her boyfriend, both open the door to her shadow, apparently the home of her raw, animal lust.

With its melodramatic flair, her story is vintage S-M. Although she obviously gravitates toward the masochistic end of the continuum, Margo is far from helpless. Not only does she tease her audience with detached assurance, she also effectively targets and draws out the sadism of her aloof customer. Consider too how she sets up her boyfriend. Why would she tell him of her slip if not to provoke him? Her confession is strategic. By stirring up his anger, she forces him to punish her by replaying the whole scenario. She ends up the eager "victim" of two angry men, yet she has transformed their hostility—and undoubtedly her own as well—into a secret source of dark passion.

Her CET is clearly concerned with mastering and choreographing her own humiliation with consummate skill. Margo provides us with only a few small clues about why she has become so brash and shrewd at creating what looks like abuse. She hints at being sexually used by at least one important man in her life, mentions her determination to be as unlike her passive mother as possible, and confesses that learning to be sexually "out there" didn't come easily. But most telling of all is her concluding comment: "Ever since I learned about my hidden sex power I've felt confident and safe."

The concept of finding pleasure and safety in humiliation is contrary to the direct validation most of us look for in sex and thus, for many, is impossible to comprehend. But Margo is far from unique. Those who find ways to satisfy their masochistic desires—perhaps by hiring a prostitute to humiliate or degrade them (it's one of the most common requests men make), convincing a lover to act out an S-M scene, or constructing a private fantasy—almost invariably report a rush of affirmation. Through sleight of hand—more accurately, sleight of mind—the skillful choreography of painful scenarios makes their sting sweet.

QUESTIONS AND ANSWERS ABOUT CETS

People are fascinated by the idea of an ultra-personal scenario underlying each individual's eroticism. Whenever I lecture about CETs I always leave plenty of time for the insightful questions and comments that invariably arise. Audiences especially want to know more about how CETs work in everyday life. This sampling of questions and answers touches on fundamental issues.

I think I understand the idea of a CET, but no matter how hard I try I can't figure mine out. Does everyone have one?

I believe each person's eroticism is shaped, to some degree, by recurring themes. But don't be concerned if you can't identify yours. Keep in mind that CETs typically operate, if not completely unconsciously, then at the edges of awareness. And there are other reasons that you might be having trouble. Many CETs are extremely subtle and easily missed. As a woman, you may have learned to focus on your partner's needs and preferences rather than on yours.

Also, women are particularly prone to feeling uncomfortable with the lusty urges expressed, unvarnished, in their CETs. Especially if you're a person who doesn't masturbate very much, or doesn't use fantasy to turn yourself on, you may have limited direct experience with the scenarios that produce strong genital responses.

There's no reason why you *have* to recognize your CET—unless you repeatedly gravitate toward sexual situations you don't like or that harm you in some way. If your sex life is fine, just enjoy it. If, however, you'd like to understand your turn-ons better, make a conscious decision to pay more attention to the images that accompany your arousal. Whatever you do, don't struggle with this process. Take your time and have some fun with it.

You seem to be saying we're slaves to our CETs. Do I have any *choice about who or what turns me on?*

Your CET isn't a prison. It's merely a framework for the creation of arousing scripts based on dilemmas from your childhood and adolescence. As your CET has evolved over the years, you've already made a number of choices that have contributed to its form and texture. Even now you regularly choose whether to embrace your CET—by trying out new variations, for example—or to fight or downplay it.

You raise a crucial question about the sphere of free will, something philosophers and psychologists will never stop debating. On the one hand, if you've ever fallen head over heels in love or been overcome by lust, you've surely noticed that these experiences happened *to* you. Your choices are limited to surrendering to or running from your desires. The only people who can honestly claim to be completely free agents in matters of eros are those who have shielded themselves from its risks—at the cost of limiting their pleasures.

In eroticism, as in life, free choice increases with consciousness. One of the major benefits of identifying your CET is that once you know what it is, how it formed, and what it's trying to accomplish, you can begin to work with it and coax it in directions consistent with where you wish to go. Those who simply act out their impulses blindly are the least free of all.

I'm turned on by all kinds of people and situations. Can a person have many different CETs?

Some people clearly respond to more than one erotic scenario. It's not unusual for someone to have a particular script for one type of partner, a completely differently kind for someone else, and still other scenarios reserved strictly for fantasy. As you know, CETs revolve around the challenges and wounds of early life. I've known people who appeared to have quite a collection of CETs, each addressing a different issue. Most of us, however, tend to zero in on one or two. One reason CETs have such power is that they are highly specific and focused.

To help you decide if you have one or many CETs, ask yourself some questions: Might there be a common thread you haven't yet recognized that subtly links the varied partners and situations that arouse you? Is it possible, for example, that your love of variety is itself a key feature of your CET? Might you be saying through your behavior, "I'm a sex fiend who can't be controlled"?

My favorite sex scene—which has to do with "servicing" someone who is completely indifferent to me except for my genitals—is obvious to me in casual sex and fantasy but seems totally irrelevant with my girlfriend. Could I have a part-time CET?

This is an extremely important point: sexual scenarios, even well-established ones, may be activated only under certain conditions. You've mentioned the three situations that are most likely to elicit noticeably different responses: fantasy, casual encounters, and intimate involvements. Like you, many men and women report that the CETs that are obvious during fantasy or in purely lusty encounters seem to disappear when they become emotionally involved.

This effect is particularly evident during early limerence, when the joy of the romantic bond becomes the ultimate aphrodisiac. Later on, however, as romantic passions cool, many find themselves missing the intensity. Some return to their CET in fantasy, either during masturbation or possibly while having sex with their lovers. But some find this awkward, especially if the flavor of the CET seems incompatible with tender feelings—which appears to be the case for you. Consider yourself fortunate that you are able to enjoy the theme of turning on an indifferent lover without having to find that lover in real life.

Sometimes I'm so horny that any halfway attractive woman will do. But if I get what you're saying, my CET should make me more selective than I am. Surely I'm not the only horny guy out there.

Rest assured you're in good company. Your CET outlines a framework for special turn-ons, not *all* turn-ons. Many people, especially men, would agree with you that their selectivity declines noticeably as horniness increases. There's no denying that some encounters are simply a means of releasing pent-up sexual tension, almost like scratching an itch. But haven't you come across women who stir qualitatively different passions in you? If you have, I bet you suddenly become far more selective, no matter how horny you are.

I used to have a thing for aggressive, cold men, but now they bore me. Did my CET change?

Perhaps. I agree with Simon and Gagnon when they say, "Few individuals, like few novelists or dramatists, wander far from the formulas of their most predictable successes."[16] On the other hand, people definitely grow. The change in your attractions away from aggressive men may represent a shift in your attitudes toward yourself and a corresponding adjustment in your CET. Or maybe you've been hurt too many times by cold men and are determined to protect yourself.

CETs normally don't change radically. I doubt, for example,

that you'll ever be drawn to passive, emotionally effusive men—although anything's possible. Most likely, you'll continue to appreciate a certain amount of aggressiveness and emotional reserve in a lover. But perhaps you now also require an all-important increase in warmth and affection. One more thing: it's certainly conceivable that you'll continue fantasizing about cold men, even though they bore you in reality. I mention this possibility because people often get upset when ingrained fantasies don't keep pace with their evolving attractions.

I'm afraid if I uncover too much about my CET I'll ruin it. Could this happen? Also, do I have to tell my spouse about my CET?

I wish I could give you the unequivocal reassurance you seek. But alas I must remind you that expanding consciousness is risky. You might, for example, be shocked by certain images that lurk, unrecognized, in the hidden recesses of your erotic mind. On the other hand, chances are slim that you will be confronted with anything you aren't ready to handle. Your mind knows when *not* to notice. Implied in your question is the realization that you can, at least to some degree, decide whether to see—and how much, how fast, how clearly, or how deeply to absorb your impressions. Experimentation is the watchword.

That said, I must also stress that expanding consciousness has a way of changing things—ultimately, I believe, for the better, but often not before shaking everything up. What if the original purpose of your CET was to convert a conflict from your early life into excitement? If you faced that conflict head-on and worked it through, you might no longer need your CET—at least not in the same way. But I doubt your CET would be totally ruined. Most likely you'd simply modify it or change how you use it.

As to telling your spouse about your CET, here I can be unequivocal: you don't have to. As we will see in Chapter 9, it may be in your interest to let a partner know something about what turns you on. But your CET itself is private. You, and only you, should decide when and if you want to discuss it.

HONORING YOUR CET

Because your CET develops gradually in response to pivotal situations, in a sense it chooses you rather than you choosing it. This means you may not feel good about everything it contains, especially any shadow material that may have become a part of it. While it is clear that CETs are influenced by cultural norms, they also refuse to be limited by propriety. By nature your CET is untamed, primitive.

For these and other reasons it is all too easy to mistrust or reject your CET. However, if you go to the trouble of uncovering it only to criticize what you discover, you'll do yourself a terrible disservice. CETs hide from the light of full consciousness to protect themselves from such judgments. If you choose to see your CET, your erotic well-being requires that you honor it. Even scenarios that appear depraved and without redeeming qualities are engaged in important emotional work and deserve your respect.

However, as we shall see in Part II, not all CETs bring joyous results. Some troublesome turn-ons prompt us to continue acting out unfulfilling or destructive patterns that may preclude intimacy or undermine self-esteem. The healing impulses of the CET can go haywire and end up hurting us in exactly the areas in which they are trying to help.

One thing you should always remember: CETs speak the primal language of the erotic mind. Learn this language, and you will know the sources of passion.

Part II

TROUBLESOME TURN-ONS

6

WHEN TURN-ONS TURN AGAINST YOU

Erotic scripts can wreak havoc by drawing you into unworkable repetitions.

The more you understand the erotic mind, the more obvious it becomes that the same depth and complexity that makes eroticism so fascinating and rewarding also guarantees that a great deal can go wrong. You know this fundamental principle of erotic life: anything that intensifies passion can also disrupt it. From the erotic equation you know that the opposite is equally true: anything that inhibits arousal can also enhance it. This means that virtually any thought, feeling, or set of circumstances can be either an aphrodisiac or an antiaphrodisiac.

In Part I you used peak experiences and fantasies as tools of self-discovery. Now in Part II you can apply your discoveries to the recognition and understanding of a variety of common erotic problems, some of which you may not have considered before. Like a lot of people, you might be reluctant to think very much about these unwelcome annoyances. They are deeply unsettling, and many people are superstitious that focusing attention on these problems will somehow make them worse.

Some people are uneasy because they have a vague awareness that their erotic patterns are troubling them but are hesitant to look too closely, especially if they're not sure what's wrong. If you sense such a reluctance within yourself, I assure you that grappling with erotic concerns directly—naming them, exploring their shape and texture, even when it's disturbing to do so—is the *only* effective way to resolve them. Long-term improvements are the rewards for short-term discomfort.

I'm absolutely convinced that if you take the time to understand erotic problems—even ones that don't affect you personally—you'll be surprised at how your appreciation of the erotic mind deepens. Eroticism is so intricately involved in the rough-and-tumble of living and loving that messy conflicts and difficulties are as unavoidable in the erotic realm as in life in general. As you know, those who expect life to be problem-free usually end up disappointed and demoralized. Healthy eroticism does not avoid problems; it works with and transforms them.

The erotic problems we'll be looking at are qualitatively different from those sex therapists have traditionally talked about. Ever since Masters and Johnson launched modern sex therapy in the late 1960s, the focus has been mostly on physiological function and dysfunction, especially two observable and measurable events: arousal (a man's erection or a woman's vaginal lubrication) and orgasm (reliably having one or more, but not "too fast" or "too slow"). As a result of this emphasis, most people assume that if the "equipment" is functioning properly everything else will pretty much take care of itself.

Sex therapy has grown, and, as with all new fields, its range of inquiry has expanded. For example, many of today's clients are concerned about a declining or absent urge for sex, traditionally called sex drive, libido, or horniness—now referred to simply as desire.[1] Neither measurable nor directly observable, desire is a totally subjective state combining biochemical influences, memories of past sex, visualizations of future possibilities, and a predilection for attending to and interpreting everyday events in an erotic way. To study desire we must move beyond our preoccupation with sex organs and venture into more elusive territory where even the most sophisticated laboratory instruments become practically useless.

In this chapter I want to call your attention to three types of erotic problems that frequently bring people into therapy. First, we'll see how some of the same emotions that intensify arousal can also produce unwanted side effects that inhibit our desire or disrupt our capacities for arousal or orgasm. Second, we'll consider how troublesome attractions can draw us toward partners who are destined to disappoint or hurt us. Third, we'll discover how love-lust conflicts sometimes make it difficult or impossible to experience affection and passion with the same person. These problems all contain a similar paradox in which long-standing and compelling turn-ons turn out to be antithetical to satisfaction.

As we've explored the dynamics of passion throughout Part I, I've given special attention to peak turn-ons anonymously described by The Group in the Sexual Excitement Survey, while encouraging you to examine your own. Although I've drawn extensively on my experience as a therapist, I believe it is crucial that our ideas about the erotic mind be based on a solid understanding of nonproblematic eroticism.

Now that we're turning our attention to troublesome turn-ons, you'll be meeting many more of the individuals and couples who have come to me for therapy. Most of what I've discovered about erotic problems and their solutions has been with the help of my clients.[2] In this chapter and the next, you'll see how a variety of people have used therapy to understand and unravel their erotic dilemmas.

As the evolution and meaning of these difficulties becomes clear, you might feel a bit frustrated when solutions aren't immediately forthcoming. Rest assured that in Chapter 8, "Winds of Change," I will describe a seven-step program that anyone can use to help with erotic problems. There you'll revisit the same clients whose quandaries you've encountered in the preceding chapters. And you'll see how they used their insights to forge new pathways to sexual healing and growth.

FEELING SIDE EFFECTS

As you know, emotions play a crucial role in virtually all memorable encounters and fantasies. The unexpected aphrodisiacs—anxiety, guilt, and anger—often have a particularly strong association with the risks and conflicts that animate so many popular erotic scenarios. Normally, these emotions cause us few if any problems because our erotic minds use them sparingly and subtly. We also employ safety factors—such as the knowledge that there is no real danger or that we are capable of handling it—to increase our comfort level.

Imagine yourself beginning a passionate encounter with your lover on the living room sofa when you notice that you forgot to close the draperies, and a nosy neighbor could be spying on you from a darkened window across the street—probably not, but how can you be sure? The idea of an unseen observer being titillated or scandalized triggers in you an exciting undercurrent of nervous uncertainty or naughty guilt. This slight nervousness enhances your arousal because you know you're actually secure. After all, the lights are low and you're not exactly parading in front of the window. No one can accuse you of being an exhibitionist. Besides, you're feeling especially proud of your sexuality tonight, and you *like* the concept of turning on or shocking a prying prude.

Keep in mind, however, that all emotional aphrodisiacs are dual edged; they have the capacity either to boost or to disrupt arousal—depending on the situation and the individual. So while you might be stimulated by thoughts of being watched, your lover might feel self-conscious until absolute privacy is restored. What for you is an excitement-boosting hint of risk is for your partner a source of genuine worry about what the unseen neighbor might think—or worse yet what he might say to others. This episode has an easy solution; you can close the drapes or lower the lights even further and continue enjoying each other. Privately, however, you may retain the image of the voyeuristic neighbor in the back of your mind as an aphrodisiac.

In this section we are concerned with men and women for whom certain emotions have become so strong and central to their arousal that they are unable to shield themselves from their disruptive effects. Typically, these are people whose erotic development took place in atmospheres thick with anxiety and guilt. Over time their erotic minds learned to transmute these feelings into erotic fuel. They became so adept at using their emotions for excitation that they stopped noticing how frightened or guilty they really were. Perhaps without realizing it they came to *require* strong doses of guilt or anxiety to become highly aroused. Unfortunately, some eventually reach the point where the same feelings that have always turned them on begin disrupting their sexual functioning or inhibiting their desire.

Brian: Relaxation and arousal don't mix

With great embarrassment Brian told me that about six months earlier he had "flunked out of sex therapy," where he had been trying to solve a distressing problem. Unpredictably and without warning his penis would go limp when he was about to start intercourse. It didn't seem to matter whether he was with Julia, his primary sex partner for almost three years, or any of the other women he sporadically dated. Although he had occasionally worried about his erections over the years, recently sex was becoming more worrisome than fun. And Julia was convinced that Brian was losing interest in her.

Their previous therapist had suggested that they work as a couple with some of sex therapy's famous comfort-building exercises. In their first few home assignments they took turns stroking each other sensuously and affectionately, bypassing the genitals to avoid performance pressures. These massages went fairly well, although Brian was painfully aware that he was not getting aroused. Before long Brian began avoiding the exercises but couldn't explain why. All he knew was that he felt completely nonsexual and was unshakably convinced that this approach couldn't possibly work. His entire being was shouting a silent but resounding "no!" to the therapy.

Brian had heard me speaking about emotional aphrodisiacs at a conference and had a strong intuition that much of what I said applied to him. "Especially," he told me over the phone, "the stuff about danger as a turn-on—that's me all over." We agreed to meet on a one-to-one basis and invite Julia to join us later, when and if that seemed appropriate. Like most people, Brian didn't talk easily about the things that aroused him. Slowly, however, as he used his peak turn-ons as avenues for self-discovery, the dramatic themes that animated his inner erotic life became apparent.

He grew up in a small, traditional Midwestern town where most people knew what everyone else was doing. As a young boy he was extremely curious about girls' bodies and was always coming up with games that allowed him to see and touch the "nasty places" that intrigued him so much. He never knew if a girl would be willing to play along, would get upset, or, worse yet, would tell on him. Worry was always a component of his sexy adventures.

Later, when a friend showed him how to masturbate, he became an enthusiast and soon also discovered the thrill of fantasies, most of which had a similar flavor of risky excitement. Meanwhile, opportunities to explore his burgeoning sexual curiosity with real live girls were few and far between. One older girl who lived on a nearby farm was a welcome exception. He fondly remembered many hours of kissing and caressing in the hayloft.

With the exception of this one older girl, all his other encounters involved long, drawn-out seductions that were both exciting and totally nerve-wracking. His strongest recollections were of trying to conceal his trembling knees and clammy hands. In short, anxiety had always been intricately interwoven with his eroticism; to be aroused *was* to be anxious.

When he left for college in a larger city, sexual opportunities were more plentiful, yet he continued to gravitate toward women who seemed reticent. With each new partner he devised clever ways to reenact the anxious anticipations that so consistently stimulated him, such as initiating sex in semipublic places or convincing his partners to try things they hadn't done before—anal sex or threeways with other women, for example. When he wasn't proposing something risky or kinky, he got his sexual juices flowing by fantasizing about one or more of his favorite anxious seductions.

He recalled several instances beginning as an adolescent when he had lost his hard-on because he was so nervous. But it would always come back—something he could no longer count on. One day Brian cut to the heart of his dilemma: "My body can't handle the danger anymore, but I'm not turned on without it." He had actually known this for quite some time but had never been able to articulate it so clearly. Now he had a way to understand why the comfort-building agenda of his previous therapy had made him so resistant. As is always the case in therapy, his resistance contained an unspoken message: "How can you expect me to solve my erection problems by relaxing when it's anxiety that turns me on?" Vetoing the therapy was a subconscious attempt to preserve the eroticism he had always known.

Brian's bind is far from unusual. Sex therapists regularly encounter men—and sometimes women too—who seem determined to avoid the comfortable, nonperformance-oriented touching that has helped thousands to resolve their sexual problems. Unfortunately, relatively few therapists are prepared to grapple with the untidy fact that anxiety isn't just a pesky impediment to be banished as efficiently as possible. The neat-and-clean view of sexuality that dominates modern sex therapy has little to offer someone like Brian.

What he needed was to have the awkward truth about his inner conflict recognized; comfort and arousal had become hopelessly incompatible. For many years Brian's CET, with its focus on the dangerous edge, had enabled him to transform anxiety into a source of excitation—well worth the occasional disruption of his sexual functioning. But now anxiety-the-aphrodisiac had become anxiety-the-erection-killer.

Once Brian understood what was going on, he invited Julia to join us for couple's therapy, an invitation she eagerly accepted.[3] As they worked together to rebuild mutually satisfying sexual interactions, Brian's attachment to anxiety as an aphrodisiac and his concern that relaxation was incompatible with arousal were an open and regular part of our discussions.

EMOTIONAL APHRODISIACS IN CONTEXT

Virtually any emotion can either promote or negate arousal. While some people are able to put up with very high levels of even the most potentially disruptive emotions, others can tolerate practically none at all. A person's response to emotion can also change over time, often dramatically, as it had for Brian. What had brought Brian's conflict with anxiety to the breaking point?

For one thing, Brian was getting older. Most men recall a time in their teens and twenties when erections were triggered reflexively by almost anything. Young men manage to get and keep erections under the most awkward conditions—such as fumbling nervously in the backseat of a car while pretending to know what they're doing. As men age, however, an increasing proportion can't always take their erections for granted. So it was for Brian, who, entering his forties, was becoming increasingly susceptible to the arousal-blocking effects of high anxiety.

Equally important was the fact that he had grown to care for Julia more than any other woman he had ever dated. Since he wanted the relationship to work, a new form of anxiety entered the picture—a fear that sex might ruin his chances for love. Also, Julia preferred easy, comfortable sex in the safety of her own bed—the polar opposite of Brian.

You can see how a combination of changing circumstances can transform an aphrodisiac into an antiaphrodisiac. And when an unworkable turn-on collapses, it often leaves a gaping void where desire used to be.

Nancy and Burt: Alcohol recovery unleashes guilt

Nancy sat quietly with her head down while her husband, Burt, summarized their problem: "Nancy and I have always had a great relationship, and that used to include sex. But ever since she—I mean both of us—decided to stop drinking fourteen months ago, Nancy hardly ever wants it anymore, which is a *big* problem for me." The room was practically pulsating with Nancy's awkward silence.

Finally, with her face contorted by a blend of shame and exasperation, she offered in a barely audible voice the explanation that Burt seemed to be demanding. "It's just that I feel so *inhibited* since I stopped drinking," she whispered. "I didn't realize how much I depended on a few drinks to loosen me up. I can hardly *stand* to have sex anymore." With the word "stand" her voice took on a noticeable authority. The lady was most decidedly turned off.

They both seemed relieved when I told them that recovering from alcoholism (or any other drug addiction) typically has significant and often unexpected effects on sexuality—even though few people discuss sex openly as part of their recovery. I must have hit a nerve because Nancy looked straight at me and proclaimed, "You got that one right. When I listen to people at AA meetings I feel like I'm the *only* one whose sex life is a total mess!" I suggested that the sexual changes they found so distressing were, in fact, a completely normal part of the recovery process. I invited them to open up the subject of sex to see what they might discover.

They had met at a party in the late 1970s. Nancy had just gotten her first job at an ad agency—a career that had flourished ever since—and Burt had been finishing a graduate degree in biochemistry. Within a few weeks they were living together. Each felt a part of the "sexual revolution" and relished sexual freedom and experimentation. Even after they married a few years later, they continued to be adventuresome by reading sexy stories to each other, acting out fantasies, and making love outdoors.

At first, a couple of glasses of wine or a few puffs of marijuana seemed to be harmless preludes to sex. Burt didn't realize there was a problem until years later when he became concerned that Nancy was drinking much more—and so was he—and that they rarely had sex without getting high.

Burt hadn't realized Nancy was never really the sexually free woman she appeared to be. Brought up in a strict family, educated in Catholic schools, encouraged by her grandmother to become a nun, she saw her sexual curiosity as horrifying to her mother and an affront to God. When she was an adolescent, sex with her first boyfriend made her both horribly guilty and incredibly excited.

As a college student in the 1970s, she embraced feminism and sexual freedom, rejected Catholicism, and found a supportive network of friends who shared her ethic of unfettered self-expression. As Nancy put it, "I thought of myself as a free thinker and rebel. Mom's disapproval made me guilty but it also inspired me." During that same period she discovered that a glass or two of wine helped relax deep-rooted inhibitions that weren't changing as readily as her ideas.

Sex with Burt was exciting and satisfying before they got married. "We were so experimental with each other," Nancy explained, "I actually *enjoyed* the guilt. I thought of us as coconspirators, saboteurs of a dying morality. We had a ball." But after they married, Nancy's strict upbringing suddenly reasserted itself. She established a closer connection with her mother and a renewed sense of loyalty to the expectations and ideals with which she had been raised. The two glasses of wine that once had calmed her inhibitions no longer did the job. The increasing physical tolerance that marks the biochemistry of addiction was abetted by a mounting psychological conflict that required ever greater doses of alcohol to quell.

Although Burt knew little of Nancy's inner struggle, he was painfully aware that sex between them was losing its spark. He interpreted Nancy's sagging desire as confirmation of his lack of attractiveness. He became so distraught that even though he worried about Nancy's drinking, he sometimes encouraged her to drink because once in a while she would let go and become her old fun self again.

Their problems escalated when they started discussing having a baby. They both knew the risks associated with drinking during pregnancy, but Nancy couldn't stop. Eventually, their marriage was wracked by drunken disagreements. The alcohol that had first entered the picture as beneficial gradually made everything worse. On the brink of separation they went to their first AA meeting together.

Burt found it relatively easy to stop drinking. Nancy too felt much better after a few weeks of sobriety. Their fighting mostly stopped, and she looked forward to the day when they could finally have a baby. Yet as her commitment to recovery took hold she felt increasingly sexless, which in turn forced her to examine her eroticism with an honesty she had never attempted before.[4]

Now that you have seen how negative emotions can cause a great deal of trouble, I hope you won't conclude—as many sex therapists and educators erroneously do—that anxiety and guilt are solely the enemies of good sex. If that were the case, eliminating the destructive side effects of these emotions would be easy because few people would be drawn to them in the first place.

However, the paradoxical perspective prepares us to accept a more complicated reality. Only by recognizing the untidy fact that emotions function as either aphrodisiacs or antiaphrodisiacs can we empathize with the struggles of people like Brian or Nancy. Only when we see the true nature of their dilemma do new possibilities become visible on the horizon.

TROUBLESOME ATTRACTIONS

Sexual attractions are among life's elemental pleasures. The simple act of noticing another or being noticed stimulates aliveness and vitality. Most people you find attractive are just passing treats for the eyes, although some become objects of longing or characters in fantasies. Not all attractions are pleasant, though. The lonely, especially those convinced of their undesirability, find the lure of a beautiful other a painful reminder of deprivation.

Whether we're looking for a casual partner or a lifelong companion, most of us defer to our attractions despite considerable evidence that they can't always be trusted. After all, attraction is often based on as little as one compelling feature, such as terrific breasts or a beguiling smile—hardly sufficient reason to pursue a potentially life-changing involvement.

The attractions that stir you aren't nearly as straightforward as they initially appear. When you feel an irresistible response to someone, your CET is probably being stimulated, although you may have no idea why this particular person is affecting you so strongly. Attractions that strike a deep inner chord do so because of a mysteriously complex and multidimensional psychic resonance. When you are strongly drawn to someone, do subtle clues and intuition allow you to perceive things about them that are normally hidden from view? Or is the object of your desire simply an appealing blank screen onto which you project a preexisting image from within yourself?

In my view, strong attractions are a baffling mixture of heightened perception, fantasy projections, and pure chance—so thoroughly intermingled that no dependable method exists for sorting them out. No wonder some attractions work out well while others are disastrous. Sometimes what you think you see is precisely what you get. At other times you may feel shocked and betrayed to realize that you are involved with someone who is not at all the person you thought. Attraction is a meaningful toss of the dice, neither a rational choice nor mere happenstance. With luck, experience has taught you valuable lessons about how to manage your attractions so that they work for rather than against you. This is perhaps the most anyone can reasonably expect.

Unfortunately, some people aren't that lucky. You've probably known people who seem to repeat the same mistakes, whether in sex or love, consistently selecting partners who aren't good for them or gravitating toward situations that turn out to be hurtful or frustrating. At first, you might think they're just having a streak of bad luck. But as similar scenarios are repeated, you naturally begin to wonder whether an unseen script directs the participants toward a predetermined conclusion. And if *you* are the person experiencing such unsatisfying repetitions, you're undoubtedly pretty discouraged.

In this section our focus is on those consistently troublesome attractions that cannot be explained by bad luck alone. Your knowledge of the erotic mind can help bring to light hidden erotic motivations with the power to excite you even as they are pulling you in directions incompatible with fulfillment. Virtually any CET contains the potential for a wonderfully gratifying involvement or for a painful replay of an old story—or a head-spinning combination of the two.

LONGING, LONG SHOTS, AND LOST CAUSES

Men and women who are sexually drawn exclusively to available partners, if they exist at all, are in a distinct minority. Who hasn't felt attracted to someone hopelessly out of reach? Some objects of longing, such as celebrities, nameless faces in magazines, or fascinating strangers admired from a distance, don't even know you exist. Others are fantasy figures borne of pure imagination. Like most people, you may have yearned for someone who was already in a relationship, too young, or uninterested in you regardless of how strongly you felt.

In the course of your lifetime you will almost certainly know the bittersweet excitement of desiring a partly available partner, happy to date or have sex with you but unwilling to get emotionally involved. If you've ever fallen under the spell of someone who vacillates between enthusiastic responsiveness and aloof detachment, you know firsthand how the intermittent reinforcement provided by an ambivalent lover makes you feel like a rat in a laboratory experiment. You press the button for a crumb of food even more frantically when the crumb is delivered occasionally and unpredictably.

Some people protect themselves from the pain of unfulfilled yearning by blunting their desire or by acting as if they're without emotional need. Others try to circumvent longing by responding only to people who pursue them, while steering clear of those they most desire. Although these strategies shield them from the pain of unrequited desire, the cost is reduced vitality and spontaneity. Luckily, most of us forge a workable, if imperfect, balance between the lure of yearning and the annoying reality that most attractions are destined *not* to be mutual.

Some people, however, seem unable to achieve such a balance. For them the experience of longing, whether or not they fully realize it, is the centerpiece of their eroticism. No longing, no turn-on—end of story. Relying on subtle signs and well-honed intuition, they are invariably attracted to those who are unavailable, confused, ambivalent, of the "wrong" sexual orientation, or who live far away. They can sniff out long shots and lost causes with amazing precision.

Yearning enthusiasts become adept at zeroing in on partners who are somewhat available, or very available part of the time, or who hint that they might become more available in the unspecified future. Longing-based involvements are passionate, stormy, and—at key moments—profoundly moving. Unfortunately, those who genuinely seek long-term relationships often discover that their commitment to longing as *the* source of excitement turns out to be incompatible with their ultimate goals. The thorniest challenge of longing-based eroticism is neither desire nor arousal—these are easy. The hard part is fulfillment.

Maggie: Master of the chase

With the wisdom that comes from tough experience, Maggie spoke a fundamental truth: "The most painful relationships to give up are the ones that never were." She was referring to a year of wrenching grief that had followed the inevitable end of a four-year affair with a married man. But on a deeper level she was summarizing a realization that, with few exceptions, her strongest attractions had always been for men whom, in one way or another, she couldn't have. Yet she dreamed of an enduring bond with someone who would desire her without reservation and enthusiastically choose to be hers exclusively. She thought she wanted a man she wouldn't have to pursue.

There was no logical explanation for her inability to find such a man. A bright, witty elementary school teacher in her mid-thirties, with a pleasing face, a shapely body, and a smile that exuded kindness, she obviously had much to offer. Over the years several men had pursued her. Yet she voiced a complaint I have heard often: "The 'normal' guys—the stable, dependable ones who would make great husbands—bore me. The exciting ones are all spoken for or on the run."

Although her latest affair had been the only one with a married man, four others had many similarities, beginning with an energetic youthfulness she found irresistible. Each also had a flair for adventure, both in and out of bed, and a knack for playful spontaneity. All her lovers had also been unreliable at times and could not be counted on to follow through with plans, return phone calls, or remember special dates—important symbols of love for Maggie.

She liked men with a rebellious streak, so, not surprisingly, all her boyfriends vigorously refused to be tied down. Yet in each she perceived a vulnerability that intrigued her. "In spite of their precious independence," she once said, "they all looked to me for stability and understanding." Maggie had always been an ultra-responsible "good girl" who always could be counted on to be cooperative, helpful, sensitive, and obedient. Consequently, although she felt miserable whenever lovers let her down, she admired their freedom to slough off so easily the obligations that dominated her.

When it came to intimacy—what Maggie said she longed for most—all her lovers vacillated between openness and avoidance. The emotional expressiveness that excited her was always counterbalanced by a fear of being possessed. "It's almost as if they see me as their mother," she mused. During her first few months of therapy Maggie eagerly theorized about the psychology of each former lover, yet it was exceedingly difficult to get her to focus on herself. As is so often the case for those in the throes of longing, the other person held the key to everything. The goal was always the same: to identify and overcome *their* hangups. In a rare moment of self-reflection she acknowledged her motivations: "If I could just win them over I would feel loved."

"And what if a man didn't need to be won?" I asked. "What if he *wanted* to love you?" After a long silence she answered, "I'm not sure I could handle it." That's when I knew Maggie was ready to direct the spotlight inward, a process that brought her face to face with her commitment to longing as her chief erotic stimulus as well as her reluctance to accept gratification.

At first, Maggie described an almost idyllic childhood. But as she relaxed her automatic tendency to make things look good, a more realistic picture emerged. She never doubted her mother's love, yet Mom usually seemed overwhelmed by responsibilities and duties. "I always had a sinking feeling that Mom was close to tears," Maggie recalled. "I knew she was terribly unhappy and it was my job to make it all better," a task at which she usually failed miserably despite her best efforts.

Even as a small girl Maggie knew that a major reason for Mom's unhappiness was that Dad worked long hours and was often on the road. When he was home he would be absorbed in the newspaper or TV. He generally rebuffed requests for companionship or conversation, whether from Maggie or her mom. Maggie later learned that her mom often suspected he was having an affair, although she never confronted him. Maggie too was unsure of her father's love, since she rarely received more than perfunctory attention from him, and hardly any of the heartfelt affection she craved.

As an adolescent Maggie found comfort in romantic stories and daydreams, most of which revolved around themes of postponed fulfillment—but always with happy endings. The imagery of yearning dovetailed with the affection-starved atmosphere of her home. In her high school boyfriend she saw the key characteristics that would consistently attract her: a devil-may-care but good-natured rebelliousness that was a natural complement to her nice-girl persona combined with a capacity for exuberant emotionality that was noticeably lacking in both her parents. For two years their relationship went well, though Maggie always wanted more closeness than what she got.

After graduation they went their separate ways. A few years later when she heard he had fallen in love and married his college sweetheart, Maggie confronted another key feature of her eroticism: a persistent undercurrent of grief and loss. Subsequent relationships all had similar outcomes—often with another woman getting, with apparent ease, the very things Maggie had struggled for unsuccessfully. Each time a man broke a date, stood her up, or withheld affection, she experienced a flood of grief, as if she were being abandoned yet again. She could only be comforted by the apologetic voice and soothing touch of her lover. When that happened her sadness would melt in a tidal wave of passion, and for a moment she would feel whole.

Her affair with the married man had been the most painful of all. From their first encounter she was convinced that he was the man she had been searching for, the man who possessed all the characteristics that fascinated her. He appeared to be crazy about her too and would often express the wish that they could be together always. He never actually promised to leave his wife, but Maggie assumed he eventually would, since he regularly spoke of a lack of love and sex in his marriage. Yet each time he chose to spend holidays or other special occasions with his wife, Maggie sank into despair. Eventually, she could stand it no longer and broke off the relationship. Unfortunately, she had been depressed and obsessed ever since. Sometimes she resorted to desperate acts such as parking in front of his house for hours waiting to catch a glimpse of him and his wife together or telephoning them and hanging up.

REPETITIONS AND REVERSALS

Maggie's story highlights one of the most puzzling questions about all troublesome attractions, whether or not they revolve around longing: Why do so many of us repeat erotic patterns that have proven to be sources of suffering and lack of fulfillment? The most obvious answer is that our attractions, no matter how troublesome, *work*. Despite the pain they ultimately cause, at critical moments they are gloriously successful at generating ecstatic passion. It is difficult to overestimate what potent reinforcers such passions are. Luckily, most of us learn from our failed relationships lessons that eventually steer us toward more workable partnerships.

Repetitive, problematic attractions are not so easily modified, which is why some people have, in recent years, labeled them love addictions. Calling Maggie's tragic search for love an addiction may provide an illusion of understanding. If, however, we genuinely wish to uncover the sources of difficult attractions we must reject easy labels and venture inward where the erotic mind has its own ways and reasons.

Maggie first encountered longing in response to the pain of not feeling loved by her father.[5] Unfulfilled cravings for affection defined the key problem her CET was trying to resolve. As her eroticism evolved she discovered—but only semiconsciously—that by continually placing herself in positions of longing she could convert her pain into passion. Logic suggests that someone in Maggie's position might be drawn to men who, unlike her father, would be naturally warm and loving. And true enough, the men she found most irresistible were capable of great warmth and tenderness, but never consistently.

To select a man who would love her wholeheartedly, Maggie would need to bypass the very dilemma her CET was designed to reverse: how to get someone who seems ambivalent to change his mind and love her without reservation. At her most passionate moments that's exactly what happened—at least for a while. Unfortunately, her CET forced her to select men who were not readily available, otherwise she couldn't anticipate the reversal that was the overriding goal of her eroticism. Her passion only sprang to life when fulfillment seemed possible but just beyond her grasp. In a moment of stunning clarity Maggie articulated the essence of her problem: "What really turns me on is *almost* being loved."

Maggie's dilemma is far from unique. Millions of people gravitate toward partners who appear to possess characteristics similar to those of significant others—parents, siblings, special friends—who have let them down in the past. Their goal is not primarily to perpetuate the pain (although this tends to be the most frequent outcome) but rather to reenact hurtful situations in hopes they can be reversed—the pain transformed into passion.

A complete understanding of Maggie's determination to avoid fulfillment requires one more piece of the puzzle, one Maggie found very difficult to accept. It's true that Maggie's first object of longing was her father, but it was through her emotional alignment with her mother that she learned on a day-to-day basis the ways of longing and sadness. Remember how desperately Maggie struggled to make her mom feel better? While it doesn't make sense logically, it is difficult for most of us to grow beyond our parents, particularly the one after whom we most model ourselves. For Maggie to receive the love she craved she had to follow a path radically different from her mother's, which seemed to her a terrible act of disloyalty.

PITFALLS OF POWER

Longing-based erotic scripts are by no means the only ones with the potential to turn against you. Regardless of the details that energize your CET, what makes your most exciting turn-ons so intense is that they soothe old wounds. Yet there's always the risk your erotic scripts will create the very problems they're trying to resolve, whether these problems originate within the family or within one's peer group.

Derrin: A need for control

Tall and lanky, with closely cropped hair and a row of pens clipped to his shirt pocket, Derrin looked just like the "technophile" he pronounced himself to be. He had a very successful career as a computer systems engineer with an income well into six figures, the respect of colleagues, and the satisfaction of regular new challenges he felt confident about tackling. He had recently been hired to teach a college course in computer science and, much to his surprise, had become a popular instructor. Students admired his soft-spoken intelligence, and some looked to him as a mentor.

Derrin wasn't accustomed to such popularity. His own life as a student had combined academic success with a chronic sense of social failure and isolation. He told me of a recurring nightmare in which he's running toward a train station to join his classmates for a field trip. In the distance everybody's joking and jostling as they board the train. He struggles to run faster but his legs feel like lead. He reaches the train just as the doors close and it pulls out. As the windows pass by, a few kids make silly faces at him but most are having too much fun to notice. "All my life," he lamented, "I've been the odd man out."

The only child of a fairly normal family, Derrin was consistently praised for his scholastic achievements. He liked school because he did well, but he also hated it because of his isolation there. With the onset of puberty he felt more than ever like the outsider. He often eavesdropped as other guys traded stories of their sexual exploits yet avoided school parties and dances, retreating instead into his studies.

One day he discovered his father's collection of sexy magazines stashed in the garage, and later in his room he pored over them. Sure he knew the basic facts about sex, but he had never felt such overwhelming sensations or such a compelling urge to touch himself. He had his first ejaculation that night—actually four or five over the course of several hours. Thereafter he developed an active fantasy life populated by lusty women eager to satisfy his every demand.

Meanwhile his relationships with real women remained distant and awkward. The ones he found most appealing wouldn't give him the time of day. But not in his fantasies. One day Derrin handed me a written copy of his most exciting one, which clearly revealed his creative formula for turning the tables:

> I'm sprawled out naked on a large bed in a dimly lit room. A big-breasted nymphomaniac is seducing me with a provocative striptease. When she removes her slip she runs it across my body. She can see I'm turned on but I conceal from her how much. After a while I pretend to be bored and turn away. Then another woman stimulates me even more aggressively until I'm on the edge of coming, still playing it cool. At the foot of my bed three women wait their turn. At the climax I throw the most attractive one down on the bed and fuck her, while the others suck and stroke my balls and ass in a wild frenzy. When I can't hold back any longer I pull out and shoot all over them.

Whereas in real life Derrin expects to be ignored by the women he desires, in fantasy the women not only aggressively break through his isolation, they actually compete for his attention, which he enjoys withholding by acting bored, until he becomes the dominant stud he always wished to be. Derrin confirmed for me that when he "shoots all over them," he's really "letting them have it"—sweet revenge for past indignities.

As we saw in Chapter 5, fantasies of dominance, often with an undercurrent of revenge similar to Derrin's, are common among both men and women. The problem for Derrin was a social life so desolate that he had few opportunities to learn about sexual interactions with real women until he entered college. There he explored dating and sex but consistently selected shy, passive women whom he felt reasonably confident he could dominate in bed even though he was attracted to sophisticated, independent women who knew how to compete in a man's world. These most alluring women intimidated him.

With his newfound popularity as a teacher he encountered a tantalizing but risky new opportunity to assume a dominant position. Against his better judgment he pursued sexual relationships with a few of the women who sought him out after class. It was enormously exciting to have access to intelligent, confident young women—the kind who had always seemed beyond his reach. The sense of importance that came with his position counteracted his usual shyness. For the first time he felt socially and sexually competent.

Also for the first time he fell in love—with a sexy sophomore named Erica. A typical date involved long discussions over dinner about philosophy, new technologies, and her career plans and personal problems. The fact that she obviously looked up to him aroused him immensely. Each date inevitably culminated in fiery sex—similar to his fantasies, but without any overt sign of revenge. "She truly admires me," Derrin explained. "I don't have to get even with her."

After a few months Erica began expressing her own ideas and opinions and deferred less to Derrin. Although he had complicated the teacher-student relationship by sexualizing it, she was nonetheless responding in her own best interests by growing stronger and more independent. She insisted on dating other guys and was offended when Derrin assumed he had a right to make demands on her. Soon the affair unraveled with plenty of hurt feelings all around. During the painful aftermath Derrin entered therapy.

The teacher had unexpectedly become the student. Now he was being forced to face some chilling realities. To begin with, deep inside he felt just as vulnerable and alone as he had in high school, and those feelings could clearly not be resolved by professional successes, no matter how impressive. Beyond that, the role of dominance in which he found solace and excitement turned out to be almost purely defensive, perfected in isolation but unworkable in real life. Only with passive women who bored him, or within the safety of a superior social position, could he sustain the illusion of control. In either case an enduring connection was impossible. As he uncovered the pain at the core of his attractions, he realized the time had come to risk a new direction.

Maggie and Derrin illustrate how few of us have as much say as we believe in choosing our attractions. We are magnetically pulled toward partners with whom we sense the opportunity to assuage a festering wound or to compensate for our incompleteness or inadequacy. Happily, sometimes that is exactly what happens. But such attractions, even as they hold out the hope of joyous solutions, also carry the possibility of drawing us backward toward repetition rather than forward toward growth. Only by recognizing this paradox can we embrace the adventure of attraction with our eyes as open as our hearts.

LOVE-LUST CONFLICTS

One of the key challenges of erotic life is to develop a comfortable interaction between our lusty urges and our desire for an affectionate bond with a lover, a bond that combines tenderness and caring with passion. In the complex world of eroticism, few people can avoid being pulled in conflicting directions, at least at times, by their dual desires for emotional closeness and raw excitement. The situations and people that stimulate us sexually may be quite different from those that make us feel affectionate. Even so, most of us learn how to accommodate our competing urges and, when the time is right, to risk the glorious roller-coaster ride when love and lust converge.

It's a sad reality that cultures and families often seem hell-bent on making the interplay of love and lust infinitely more difficult than it needs to be. The framework for adult love-lust interactions is sketched out during childhood and adolescence. Adult lust evolves out of simple experiences of sexual and sensual curiosity. The precursors to adult love are early emotional attachments, first to Mom, then to the larger family, and eventually to a community of friends and acquaintances. When kids play house, mimic favorite TV characters, simulate flirtatious poses or embraces, or initiate or reject nudity or sexy games with their peers, they're engaging in the important work that Dr. John Money calls "sexual rehearsal play."[6] Through positive sex play kids learn about their own and others' bodies. At the same time, the complex interactions and negotiations required by their games provide opportunities for practicing skills of social conduct.

When adults try to protect children by forbidding sex play with age-mates, the message is powerfully communicated that their curiosities are bad, setting the stage for a later mistrust of lust. These "protected" children are frequently unprepared to handle the more demanding interpersonal challenges of adolescence. As adults they often end up confused or in conflict about matters of love and lust.[7]

For those who hold the twin beliefs that love is good and lust is bad, two basic strategies are available for dealing with this dichotomy. One approach seeks to tame or to purify lust by fusing it with love so that lusty urges are interpreted as love, an especially common response to sex-negative training among women. Rushes of sensation that many would call lust are interpreted as love or affection. A love-lust fusion can be visualized like this:

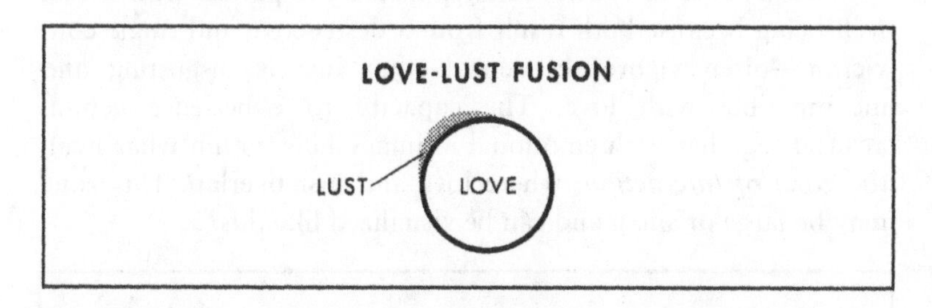

When sex must always be joined with love, even casual sex can become infused with emotion and meaning. Unfortunately, this special intensity is typically followed by the realization that the partner isn't feeling the same way. More often than not, the dream of lust redeemed by love quickly turns to loneliness. The inability to experience lust as separate from love clouds one's judgment and is responsible for incalculable unnecessary suffering.

Another response to love-lust conflicts, more common among men, is to protect the purity of love by creating an invisible barrier between it and lust so that a person becomes unable to feel both at the same time or with the same person. The splitting apart of love and lust can be visualized like this:

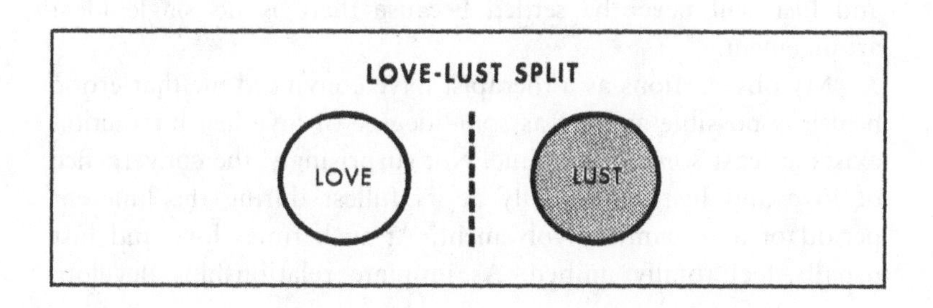

When someone pursues lustful aims with complete detachment from tenderness and affection, erotic attention narrows to a laser-like focus on maximum genital arousal. The result is often a level of excitation that is qualitatively different than any other kind—hotter, more insistent, a unique psychophysical high. Unfortunately, love-lust splits make it difficult or impossible to build and maintain sexual relationships because affection intrudes, reducing or obliterating the sought-after intensity.

Neither love-lust fusions nor splits are compatible with sexual well-being because both result from a destructive and tragic conviction—often vigorously denied—that lust is disgusting and incompatible with love. The capacity to experience genital arousal together with emotional intimacy falls within what I call the *zone of interaction,* where love and lust overlap. This zone may be large or small and can be visualized like this:

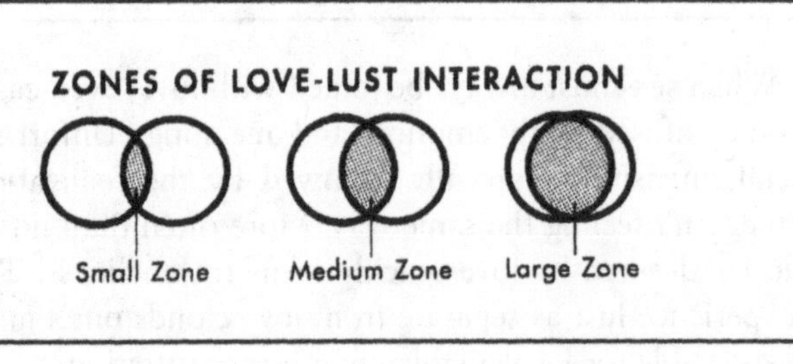

Disagreements abound regarding the optimum amount of love-lust overlap. There are those who strongly believe, as a matter of morality or preference, that eros reaches its full potential only to the degree that love and lust are experienced in tandem. Others feel especially alive and vital during experiences of uncomplicated lust. Disputes about the best relationship between love and lust will never be settled because there is no single ideal arrangement.

My observations as a therapist have convinced me that erotic health is possible as long as *some* degree of love-lust interaction exists at least *some* of the time. Not surprisingly, the convergence of love and lust is normally at its fullest during the limerent period of a romantic involvement. At such times love and lust usually feel totally unified. As intimate relationships develop, however, the zone of interaction normally grows smaller and less consistent. Virtually all long-term couples grapple with this change, and hardly any are pleased about it. In most cases, however, as long as affection and lust continue to interact to some degree, the partners stand a good chance of finding ways to continue enjoying sex with each other.

PRISONERS OF PROHIBITION

We have devoted considerable attention to the naughtiness factor and the way it brings the thrill of the forbidden to all sorts of encounters and fantasies. You've probably noticed in your own life how a sense of raunchy fun can add a welcome spark to already satisfying sex. The ability to transform naughtiness into arousal begins as a creative adaptation to the distressing fact that the adults upon whom we depend for survival don't approve of our sexual curiosity.

When you have fun with naughtiness, you acknowledge the restrictions you faced as a child while asserting that, to a significant degree, your desires have triumphed over the forces that tried to suppress you. You have grown sufficiently free to use prohibitions for your own enjoyment and play. Many people, however, grew up in homes so permeated with antisexual restrictions that the drama of violating prohibitions has become the central feature of their eroticism. These people are prisoners of sexual prohibition, and it's no exaggeration to say that they have been victimized just as surely as if they had been molested.

For prisoners of prohibition every opportunity for enjoyment sets off a frantic inner conflict. When they feel aroused they're awash with guilt. If their inhibitions are temporarily swept aside by desire, they pay later with remorse and shame. This conflict is so unpleasant that some people develop sexual aversions and go to great lengths to avoid becoming aroused. In the worst cases perfectly normal sexual stirrings actually trigger panic attacks.

The opposite reaction takes place in those who become obsessed with sex, compulsively reenacting the very behaviors that were most forbidden. They keep their eroticism alive by sustaining a hidden sex life, cut off and thus protected from antisexual judgments. If these secret sex lives are eventually exposed, it is often to the amazement of shocked friends, family, constituents, or congregations.

Many sex offenders are prisoners of prohibition who can only become aroused when they are actually violating laws, risking legal punishment and social censure. But most prisoners of prohibition are not sex offenders—and never will be. Instead the perpetual tension between antierotic training and persistent forbidden desires fuels an internal struggle as compelling as it is grueling. They harm no one except themselves and those who try to love them. In most cases their torturous secret comes fully to light only when they try to form an enduring sexual bond.

Ryan and Janet: Too close for lust

Ryan was distraught over his relationship with Janet, his new girlfriend. He told me how important she was to him, how doubtful he was about his ability to succeed at romance, and how determined he was to make a go of it. Ryan's relationship with Janet was the most affectionate and tender he had ever known. "But," he continued with a deep sigh, "I'm not handling it very well." He was embarrassed to admit that even though he found her very attractive, he had never been able to ejaculate with Janet. As we discussed his predicament it soon became apparent that his inhibited ejaculations were a manifestation of a much more serious problem: a gaping chasm between his lusty urges and his desire for love and affection.

I asked him to recall his earliest feeling or experience that seemed, in retrospect, to have been sexual. Most clients need to think about this question for a while, but Ryan knew his answer instantly. Around age four or five he had been casually playing with himself, enjoying the warm, tingling sensations between his legs. This simple memory became permanently etched in his mind when his father walked in on Ryan's sensuous reverie, flew into a rage, ranted about hell and evil, and commanded him never *ever* to do that again, terrifying little Ryan half to death.

"From that moment on," Ryan proclaimed, "I knew I *must* restrain myself. Soon I also knew I was obsessed with my little prick, how good it felt, and how bad I was for wanting to touch it." He went on to explain that, with plenty of twists and variations, he still felt essentially the same way as a middle-aged man. He craved love yet was driven by an insatiable appetite for nasty sex. In the past he had often hired prostitutes but not since meeting Janet. Thus far, though, he had been unable to resist even stronger urges to masturbate while talking on telephone sex lines. And he still sneaked out to X-rated bookstores in the black of night in search of pornographic videos and magazines. He was fascinated by graphic depictions of down-and-dirty sleaze.

He was the first person I ever met who called himself a "sex addict," at least ten years before the term became fashionable. The connotations of addiction accurately captured the out of control, compulsive quality of his behavior. But his insistence on casting himself as an addict who had to "kick the habit" only intensified the grueling struggle that fueled his obsession. Like most of today's "sex addicts" Ryan was less hooked on sex than on *fighting* with it. He felt the helplessness of a drug addict because for him forbidden desires escalated in tandem with the need to resist them. His excitation was always ultra-hot, but there was little or no lasting enjoyment. For as long as he could remember, he had most assuredly been a prisoner of prohibition.

No wonder Ryan was easily orgasmic while masturbating to porn or with anonymous phone partners. He considered his fantasy women sluts with whom he could freely express his depraved, lusty self. The fact that his fantasy partners weren't actually present also made it easier for him to express his impulses in unedited form, although he worried that one day someone would recognize his voice and expose his secret. Because he admired and respected Janet, he felt self-conscious and inhibited—cut off from the roots of his eroticism. Although their shared affection and sensuality felt marvelous and usually produced an erection, he was unable to surrender to lust and generate sufficient erotic intensity to push himself "over the top" to orgasm.

Once Ryan understood his conflict he asked Janet to join him in couple's therapy. He wisely decided that telling Janet about his lusty exploits outside their relationship would only hurt her, so he kept those details to himself. But he was determined to tell

Janet about the inner conflict that was preventing him from feeling sexy with the woman he loved.

ECHOES OF OEDIPUS

Since I already had a relationship with Ryan, Janet and I had a few individual sessions. She felt great admiration for Ryan, whom she saw as talented, highly principled, and committed to making the world better. Janet felt frightened because she had previously loved two other men whom she had also admired. Both relationships ultimately failed, largely due to sexual incompatibilities. She lived in dread that history was about to repeat itself.

Janet had always been attracted to kind, intelligent men, beginning with her father, whom she adored. Her mother, whom she also loved, became terminally ill when Janet was eleven and was bedridden for almost two years before she died. Although her two brothers helped a little, Janet and her dad did most of the caring for her mom and kept the household going. Her father's unwavering commitment to the family deepened Janet's admiration. The two of them often cried together, comforted each other, and shared their hopes and fears.

Once we began couple's work, Janet was very understanding when Ryan told her how he had learned from his father that sex was sinful. She was surprised to hear about Ryan's current battle with lust but also relieved that Ryan's troubles weren't her fault. Together they agreed to venture into the sexual realm slowly through sensual touch and open communication. But before long Janet felt uncomfortable, especially as their touching expanded to include each other's genitals. In an unexpected reversal of roles, the more at ease Ryan became with sexual contact, the more Janet avoided it.

During one pivotal session Janet confessed that sex with Ryan felt "almost incestuous." Ryan looked shocked as she explained, "You and I are so close, sometimes I think of you more like a brother than a lover." Very gingerly they opened a topic shrouded in the most powerful of all taboos: sexual feelings within families. Janet gradually realized that her incest fears were an unresolved aspect of her relationship with her father. During her mother's illness and after her death, Janet had clung to him. But she also recognized, with considerable guilt and anguish, that his reliance on her as helpmate and confidante had felt "a little weird." She explained, "I think maybe we became *too* close—not that anything sexual ever happened between us—just emotionally. I remember thinking, 'Now I have to take care of Daddy,' as if I were supposed to be his new wife." She shuddered noticeably as she added, "I can't believe I said that."

This all took place as she was entering puberty. It surprised her that she couldn't remember any sexual feelings at all during her adolescence. Clearly, she had unconsciously suppressed her sexuality to avoid complicating her confusing emotional bond with her father even further. Once she started dating, she typically formed platonic friendships with men. In three amazing feats of intuitive attraction she had selected boyfriends with whom she could be close but who, because of their own insecurities, were reluctant to have sex with her. Thus she had spared herself the daunting task of examining her incest fear, the one difficult aspect of an otherwise wonderful relationship with her father. By keeping the spotlight on her boyfriends' sexuality rather than on her own, she had kept her own lust in check.

During our next session Ryan revealed that he often felt like his mother's "little husband." That day Ryan had been talking with his mom on the phone when she confided in him, as she often did, how lonely and unhappy she was with his father, who didn't seem to care that much about her. As usual, Ryan was trying to comfort her when she whispered, "Oh, Ryan, you're the only one who understands." Although the exchange was all too familiar, a wave of revulsion suddenly engulfed his body, along with an overwhelming urge to hang up. Ryan added, "Sometimes she's so damn needy it gives me the creeps."

Almost everyone is aware of how damaging sexual contact with an adult can be for a child. Not so widely recognized, however, is that certain kinds of overclose emotional involvements between parent and child, even when no overt sex is involved, can make it very difficult for the child to integrate love and lust as an adult.

When Ryan and Janet made love, the combination of their intimate connection with sexual arousal apparently triggered old incest fears in both of them. When Ryan engaged in phone sex with sluttish fantasy women, not only was he acting out his identity as a depraved sex fiend, he was also directing his erotic attention toward someone as unlike his mother as possible. By separating love from lust he was honoring his father's warnings while also steering clear of any interactions that might feel even vaguely incestuous.

Love and lust are inseparable parts of a larger whole for some, while for others they are irretrievably disconnected. Most of us, however, express our eroticism somewhere in the gray areas where love and lust both relate and conflict. Only by realizing that the two experiences are separate can we avoid painful self-delusions; to lust is not necessarily to love. By also recognizing that love and lust can and do interact, we open the door to the deepest mysteries of eros.

COPING WITH TROUBLESOME TURN-ONS

Each individual's CET evolves in response to the challenges and conflicts of early life. The erotic mind attempts to gain mastery over these problems by using the obstacles they present to stimulate desire and arousal. You can see how everyone in this chapter except Janet had learned to use long-standing conflicts quite successfully as turn-ons. But while most found plenty of excitation, they also discovered that their erotic scripts ultimately perpetuated the very problems they were trying to resolve.

If you identify with any of the troublesome turn-ons described in this chapter, the best thing you can do is to use the understanding you have gained thus far to help you recognize how your patterns of arousal are working against you. Without this awareness you must blindly follow the established path wherever it leads you.

Now is a good time to conduct a simple reassessment of your eroticism to see if you might be affected by any of the three types of erotic problems we've explored in this chapter:

Feeling side effects

Troublesome attractions

Love-lust conflicts

First, reconsider the emotions most prominent in your CET. You already realize that the full range of human emotions can energize your turn-ons and that some feelings—especially anxiety and guilt—can disrupt your body's ability to function sexually. Consequently, if you're grappling with a sexual dysfunction, pay careful attention to the emotional aspects of your CET. Keep in mind that, like Nancy and Burt, you may not easily recognize how guilty you actually feel, especially if you regularly rely on alcohol or other drugs to calm your inhibitions. Or, like Brian, you may have become so accustomed to anxiety in your sex life that you hardly even notice it.

If you've been unlucky in a string of relationships, could it be that you are reenacting frustrating or painful relationships from the past? Make a list of each of the people to whom you've been most strongly attracted in your life. What characteristics do they have in common? Can you identify difficulties from your early life that you're still trying to fix or otherwise play out through your current relationships? Unresolved feelings from historically significant relationships, particularly those within the family, are natural aspects of attraction, so there's no need to feel ashamed if you recognize them within yourself. The important thing is to notice if you keep trying to reverse a painful relationship from the past by unwittingly selecting partners with whom you are doomed to repeat it. This dilemma can only be resolved when you are willing to bring your motivations into consciousness.

Finally, consider the degree of connection between love and lust in your eroticism. It helps to sketch out your personal zone of love-lust interaction. As I demonstrated earlier in this section, draw or visualize two circles, one representing lust and the other representing love. Position the circles to show how much they've overlapped in your past as well as in the present. Like many people, you may have developed two parallel systems of attraction, one focusing on qualities you associate with sexual magnetism and the other emphasizing features compatible with intimacy and affection. Unless these sets of attractions allow for at least some degree of overlap, you'll probably have trouble maintaining a long-term sexual relationship. On the other hand, if you find it difficult to distinguish between love and lust, you're especially vulnerable to unworkable involvements. By recognizing where you stand, you can identify potential problems and prepare for constructive solutions.

Once you've acknowledged a troublesome turn-on, what should you do about it? In many instances confronting your predicament creates conditions ripe for growth. Luckily, the erotic mind is quite capable of making positive adjustments based on lessons learned from experience. In Chapter 8, "Winds of Change," we'll focus on how to nurture the creative, adaptable characteristics of your eroticism. In the meantime, nonjudgmental self-observation will help you reclaim the ability to choose.

7

SEX AND SELF-HATE

When low self-esteem fuses with high arousal, the results are the most destructive of all turn-ons.

The infinitely variable scripts that shape and energize erotic life share a common goal: to affirm the value and desirability of the story's star. But what if your eroticism began evolving at a time in your life when persistent feelings of inferiority were dictating your self-image? What if profoundly negative beliefs about yourself had become so thoroughly woven into your turn-ons that they simultaneously excited you immensely and made you feel worthless? To complicate matters still further, what if you were only vaguely aware that all this had ever happened?

Chances are that you or someone you care about is struggling with this most perplexing and troublesome of all turn-ons: the fusion of sexual excitation with self-hate. Although hardly anyone discusses them openly, erotic problems linked to low self-esteem are distressingly widespread and are the focus of this chapter.

You might assume that such profound conflicts would be easy to spot, but frequently they go unrecognized. For one thing, the symptoms of eroticized self-hate are amazingly variable. They can include chronic mood disturbances, difficulties becoming aroused or having orgasms, an absence of sexual desire, or an overwhelming uptightness or aversion toward sex. Some sufferers derive unmatched intensity from compulsively reenacting their grueling inner struggle in nonstop sexual repetitions that generate enormous heat, but little satisfaction.

CORE BELIEFS AND EROTIC DEVELOPMENT

To understand how excitement and self-hatred can become intertwined, it is necessary to consider the role of deeply held, longstanding beliefs in shaping your self-image. Core beliefs are thought patterns that take root in the first few years of life and become firmly established as you grow up. You might think of them as a summary of the key convictions you hold about crucial issues such as the kind of person you are and will become, your place within your family and the world at large, and how you expect others to respond to you. Core beliefs are so thoroughly integrated into your personality that instead of thinking about them directly, you simply experience them as givens.

Not nearly enough attention has been devoted to the role of core beliefs in shaping eroticism. Your most potent and enduring core beliefs are the raw psychic materials with which you subconsciously construct your unique blueprint for arousal. If you are lucky enough to grow up believing that you're basically an acceptable person who, in spite of imperfections and flaws, is worthy of respect and love, you will gravitate mostly toward partners and situations that directly or indirectly affirm your worth and desirability.

Of course, no matter how much love and validation you received as a child, you must still grapple with doubts about your value and adequacy. Your erotic mind is keenly aware of your perceived limitations and vulnerabilities, and tries to compensate for them by transforming self-doubt into validation, weakness into power, pain into pleasure, and everyday trials and tribulations into erotic possibilities. No matter what internal or external obstacles stand in the way, the erotic impulse springs from a deep-seated urge to affirm yourself. One thing, however, is absolutely required for your erotic mind to function positively: deep within your psyche at least a small seed of self-worth must have been planted and nurtured.

Unfortunately, a wide variety of childhood and adolescent adversities can foster such intensely negative core beliefs that the natural urge toward self-affirmation is squelched or distorted. Sometimes, for example, a chronic illness or disability, especially if accompanied by severe teasing or ridicule, can lead a child to conclude that he or she is simply too defective ever to be acceptable.

If you've ever observed children closely you've probably noticed tremendous variation in how they respond to the inevitable challenges to their self-esteem. Some are unable to shield themselves from the slightest criticism, while others appear relatively unruffled by even ruthless put-downs. Personality differences account for some of these variations, but it is a great source of security if at least one parental figure unequivocally and dependably reaffirms the child's worth. Children who consistently feel nurtured, protected, and encouraged permanently internalize these parental functions as parts of themselves. The memory of having been loved and supported remains a reliable resource, a shield against the harsh realities of life.

TURNING AGAINST ONESELF

Instead of being nurtured and protected by those they depend upon for survival, many children are treated with indifference or contempt. All forms of neglect and abuse have similar effects. One of its most devastating subconscious legacies is a tendency for the child to assume that he or she *deserves* to be treated badly. Such a conclusion is clearly not based on adult logic. Yet most children will readily sacrifice their own self-esteem to justify their parents' attitudes and behaviors toward them.

More often than not, abused kids interpret their mistreatment as proof that they are fundamentally and irreparably flawed. They feel a profound loyalty to their abusers and consequently absorb intensely negative beliefs at the deepest possible level. Some become harshly self-critical and perfectionistic, forever gathering evidence of their shortcomings. Others are obsessed with winning approval by trying to please everyone. Some become superachievers, but no matter how much they accomplish, the inner conviction of worthlessness holds firm. More than a few are determined to demonstrate, over and over, the accuracy of their negative core beliefs by behaving in ways that invite the very things that have been most hurtful: punishment, rejection, and defeat.

When negative core beliefs conflict with the erotic mind's search for validation, the self—wanting to express its potential, yet convinced of its unworthiness—is torn by a profound inner struggle. Some people manage this conflict by allowing themselves to thrive in one or two areas, while acting out their self-contempt in others. I've known some people who were their most self-affirming in the erotic realm. They enjoyed a rewarding sexual life but deprived themselves of most other satisfactions. Such people are exceptions, however, because eroticism is particularly sensitive to the damaging effects of self-hate. It's not unusual for a person's capacity for enjoyment to break down completely under the crushing weight of this relentless inner conflict.

In most cases self-hatred doesn't obliterate one's sexual impulses. Instead the erotic mind rises to the task of devising *indirect* routes to self-affirmation. A person's CET can meet this challenge in many ways, but there are two fundamental strategies. In the first, the self-hater becomes the master of his or her suffering by actively seeking out and choreographing sexual situations that are ultimately demeaning. The second strategy also creates an illusion of control when the self-hater inflicts on others indignities similar to those he or she has suffered; the victim becomes the victimizer. In my experience, CETs driven by self-hatred usually combine elements of both strategies in a frenetic search for validation.

Regina: Wounded seductress

With a dull serrated kitchen knife, Regina had etched a crosshatch of jagged cuts on her left arm and wrist. Her roommate found Regina sitting at the kitchen table, still holding the knife, with blood oozing in zigzag trails down to her fingertips and tears streaming down her face. Doctors at the emergency room said she was lucky; the cuts were superficial. Regina insisted that despite appearances she wasn't really trying to kill herself. Yet she was unable to give an alternative explanation for her behavior.

Those who knew her were amazed. A twenty-two-year-old college junior with decent grades, an active social life, and a small circle of close friends, she seemed perfectly normal. But now that the enormity of her unspoken distress was out in the open, she was eager for help. The next day she began individual therapy with me and soon joined a women's group as well.

During our first meeting she stared vacantly into space and spoke in a flat, emotionless voice—not surprising, considering her ordeal of the previous day. What did surprise me was that she was wearing a see-through blouse, and that even in her trancelike state her postures and movements seemed calculated for maximum seductive effect. I learned that Regina's mother, a nurse who was struggling to put her through college, had caught the first plane to be with her. I also learned that her father had died when she was a little girl, that her mother had remarried when Regina was four, and had divorced her second husband when Regina was a teenager. "What a jerk!" was all she would say about her stepfather, but she declared it with bitter conviction.

She was truly dumbfounded by her self-mutilation. She had imagined cutting herself before but never seriously thought she'd go through with it. I explained that hurting oneself physically is often a means of dulling a more severe emotional pain. She nodded but veered off in a new direction. "The strange thing is," she said, "I've been doing pretty good lately. I'm dating this really nice guy who treats me like a lady." She explained that most of her boyfriends had wanted her only for sex. "Not that I didn't want it too," she added, assuming an even more seductive pose. I was being tested.

During our next meeting I invited her to use our sessions to discuss whatever was important to her, including sexuality. I emphasized that she and I would only meet during scheduled therapy sessions and never have any form of sexual contact. Almost immediately she shifted to a more natural posture and allowed her eyes to meet mine comfortably for the first time.

A few weeks later she mentioned, almost in passing, that when she was a girl her stepfather had "fooled around" with her, but she immediately slipped into icy silence. As long as we didn't discuss her stepfather she was eager to tell her story. She had been one of the first girls in her class to develop breasts, which she quickly discovered were of great interest, especially to older guys. By age twelve she was sexually active. Her "specialty" was seducing men, which made her feel powerful, desirable, wanted. Yet she had never loved or felt loved by any of her dozens of lovers.

She confessed that being held and caressed meant much more to her than sex, yet most guys she dated hadn't been very affectionate. Later she confided to me that intercourse was sometimes painful and she often tried to avoid it. "I get guys off before they have a chance to fuck me," she said. She believed that her discomfort with intercourse was a major reason that so many men had rejected her. "Fucking is all they care about," she complained with a look of deep sadness. "Sometimes my vagina clamps down so tight it hurts," she explained, "but I just can't help it." She was relieved to know that her problem had a name—vaginismus—and that it happened to many women. One of the things she appreciated most about her current boyfriend, Bill, was that he seemed to crave affection as much as she. He also sensed her discomfort with intercourse and respected it.

As she became more comfortable with me her references to being sexually molested by her stepfather increased. She continued to grow distant, though, whenever I asked her to elaborate. "I'm not even sure *what* happened," she would often say. "My memory is so fuzzy. Maybe I'm just making a big deal out of nothing."

Unlocking memories of childhood abuse is torturous, delicate work. Children molested by those on whom they depend will go to great lengths to conceal, even from themselves, the full truth of what they have endured. During abusive episodes, boys and girls typically protect themselves by going into a trancelike state called dissociation.[1] Consequently, they aren't fully present during their abuse, which, of course, is a blessing at the moment. As time passes, painful memories are kept in check with an arsenal of defense mechanisms, especially repression (banishing painful memories from consciousness), denial (acting as if it didn't happen), downplaying (minimizing the destructiveness of the experience), or rationalizing (concocting plausible explanations for the abuser's behavior). So much effort goes into not experiencing the pain, no wonder unambiguous memories are difficult to grasp.

Yet traumatic secrets crave the light of day almost as much as they shrink from it. Years or decades after the original trauma, particularly when the person's psychological resources have grown stronger, inner pressure builds to reclaim the truth, no matter how painful. Some may be taunted by visual images that flash in and out of consciousness. For others, certain smells, sounds, tastes, or emotions are memory catalysts.[2] Sometimes the subconscious uses the symbolic language of dreams to nudge necessary but unwanted material closer to awareness. Nowadays public disclosures by celebrities and others also serve as memory triggers.

In the midst of our therapy, Regina's memory was jolted while she was watching a group of women on "The Oprah Winfrey Show" who had been abused as children. As each guest told her story Regina felt sorry for her yet strangely detached. "I was wondering what to make for dinner or some such nonsense," she explained, "when suddenly a woman's voice grabbed me as if someone cranked up the volume full blast. She was complaining about how her mother had refused to listen when she tried to tell her that a family friend was putting his hand inside her panties after school. I became instantly *enraged* and shouted at the top of my lungs, 'Believe her, goddamn you!'"

Over a period of many months, Regina told me, the members of her group, and her closest friends the truth she had not wanted to know but always had: the man she had called Daddy had seen her as little more than a sex object. Beginning when Regina was six or seven he would slip into her room when her mother was nursing on the night shift. For a few moments he would stroke her hair and tell her in a soothing voice how sweet she was. She loved his affection, his strength, and the warmth of his caress; the attention made her feel special. But she soon learned to associate these good feelings with the terror that would inevitably follow.

He would remove her pajamas and his own, and his gentle touches would turn cold and hard. Instead of caressing her, he would grope her roughly, including between her legs, jabbing his penis against her. She wanted to scream out but instead would repeat in her mind, "Daddy loves me, Daddy loves me," over and over and over until the pain subsided. She would "go away" until she fell asleep. When she awoke the next morning she could barely remember anything except for the aching shame that was with her always. These episodes continued until she was about twelve, when she began protesting more vigorously. Perhaps he feared she would tell, or perhaps as she entered puberty she became less interesting as a sex object, or perhaps he found someone else to molest. Whatever stopped the abuse, shortly thereafter she became a seductress—not an altogether surprising development.

Although she had never mentioned her writing abilities before, one day Regina came in clutching a stack of stories she had written long ago. I asked if she would pick one and read it aloud After a few deep breaths, and in a strong, clear voice, she told the "fiction" of a little girl under siege and the imaginary friends who had been her witnesses. My first urge was to distance myself from the overwhelming emotions—terror, rage, utter desperation—by commenting on the impressive writing, but I remained silent. I was simultaneously shocked, outraged at this brutal man, and in awe of Regina, whose creative impulse had helped her to preserve the truth and survive the intolerable.

She was equally courageous about facing what she came to call "The Big Lie"—the belief that her stepfather's behavior had something to do with her being special. But face it she did, and with her realization came a flood of tears that seemed at times as if they would never stop. She found it almost as difficult to acknowledge the anger she had felt toward her mother, the one whose love she had always counted on but who, for some reason, had been unwilling or unable to protect her. When Regina found the courage to bring it up, her mother confessed her guilt as the two held each other and sobbed. Yes, she had suspected something wasn't right. Yes, she remembered Regina begging her not to go to work. But she just couldn't deal with it.

As she reclaimed her painful history, Regina began reexamining her eroticism with a piercing new clarity. As a result of her stepfather's abusive actions, she had developed a core belief that she had value only as a sex object. That was the one role in which she had a ghost of a chance of receiving even a speck of affection and tenderness from him. Her CET expressed these self-deprecating beliefs at the same time that it provided her with an indirect mechanism for affirming and protecting herself. Each time she invited—no, *compelled*—men to use her for sex, she became the master of her fate. At the same time, her vaginal muscles formed an involuntary barrier that expressed the "no" she couldn't speak.

The twin themes—repetition and reversal—that were so prominent among the men and women you met in the last chapter are clearly visible once again, only more dramatically here. By continually reenacting her trauma with a new twist, Regina symbolically converted her exploitation into a source of erotic power. As the manipulator of men's sexual urges she shielded herself from the intolerable truth that Daddy had treated her as he had because he didn't give a damn about her feelings. "I'd rather be the slut who seduced my father," she once proclaimed, shocking herself with the implications of her insight, "than the worthless piece of shit he used."

In Regina both the power and the vulnerability of the human psyche are clear. Her CET was a stroke of genius through which she nurtured a kernel of self-respect. Unfortunately, because her eroticism revolved around the belief that it was her place to be exploited by men, her heroic attempts at self-affirmation continually brought her back to feeling used for sex and then cast aside. But then, with no conscious idea of what she was doing, she had made a suicidal gesture and thus initiated a series of events that ultimately exposed her inner wound and unlocked the healing power of the truth.

WERE YOU ABUSED?

As you can see, the most profoundly damaging core beliefs develop in response to severe abuse—emotional, physical, or sexual—as a child or adolescent. For healing to begin, it is necessary to piece together the details of exactly what took place, and then to tell the unvarnished truth to at least one other person.

Telling one's story solidifies its reality. Emotional abuse, such as constant demeaning put-downs, is more easily denied than sexual abuse. I've known people who were regularly threatened, chastised, or even severely beaten yet believed that these assaults were merely discipline. Growing up in such an environment makes it difficult to know what "normal" is.

Some people try to forget about terrible childhoods and move on. They fail to recognize that to move beyond a trauma they must come to terms with it, which is impossible if the facts remain a blur. Once you know what happened and how it affected your beliefs and expectations, you can claim responsibility for your present choices, repair some of the damage to your self-esteem, and learn how to give yourself the respect and nurturance you deserve.

There are, however, potential perils involved in letting out memories of past abuse. As memories grow clearer some people become overwhelmed by depression, fear, rage, or despair. For the first time they feel the full impact of the self-loathing that is a byproduct of their mistreatment. Some even become suicidal. That's why it's crucial to have the support of friends and loved ones, and perhaps professional assistance as well.

Another danger is that the person may find a paradoxical sense of comfort or meaning in the role of victim. More than a few people cling to the belief that they must remain forever helpless and therefore unwittingly perpetuate their own abuse. They continue to suffer needlessly until they realize that victimization is not an identity to be embraced but a harsh legacy to be recalled and overcome.

Finally, some people have false memories, although I hesitate to say so.[3] After all, most people will go to great lengths to downplay abusive experiences. The last thing they need is an additional worry that they might be fabricating or exaggerating. Most men and women who were abused as children have at least vague memories even if they have been clouded by years of denial and rationalization. Usually, it's a matter of bringing to light something the person already suspects or knows.

But beware of anyone, especially a therapist or counselor, who tries to *convince* you that you were abused—especially if you never thought of it before. Fragile memories can easily be suppressed, but they can also be distorted or embellished by those who are overeager to see abuse as the cause of all problems and unhappiness. Never let anyone impose his or her intuitions on you, for this is just another type of violation. You must be free to discover your own truth.

TRAUMATIC APHRODISIACS

It's easy to see how abusive experiences that foster self-hate can disrupt someone's ability to experience intimate touching as pleasurable and result in aversive feelings toward some or all forms of sex. Whereas Regina's aversion focused particularly on intercourse, for others all sensuous touch is so thoroughly contaminated by negativity that they are unable to make any meaningful distinction between pleasure and pain. To them, *pleasure* hurts—so they vigorously avoid it. For reasons that aren't well understood, women are more likely than men to develop sexual aversions in response to trauma, although some abused men also become pleasure-phobic.

Contrary to the rules of logic and common sense, some cope with intolerable pain by transforming it into the most irresistible of all aphrodisiacs. This is possible because of one of the most curious oddities of the erotic mind. On the razor's edge between affirmation and degradation, beyond the usual categories of pleasure and pain, there exists an overpowering form of emotional and physical excitation that I call *pleasure-pain.*

I'm not talking about nonproblematic forms of intense stimulation—slapping, pinching, or biting, for instance—that passionate lovers or S-M enthusiasts frequently enjoy at the boundary of pleasure and pain. Such experiences can be completely compatible with a fulfilling sex life. Our concern here is with a qualitatively different type of experience in which pleasure and pain have become so thoroughly fused together that sensations, thoughts, and feelings that would normally be experienced as highly noxious instead become inexhaustible sources of erotic fuel.[4]

THE PUZZLE OF PARAPHILIAS

The effects of pleasure-pain are most dramatically expressed in paraphilias (from the Greek, meaning "outside of love"). These are "kinky" sex rituals that so thoroughly take over some people's eroticism they are unable to reach orgasm without acting out or fantasizing a highly specific scenario. A paraphilia is always intensely exciting, but rarely much fun. It is the ultimate manifestation of pure, focused lust in its most grueling and compulsive form.[5]

Extreme forms of voyeurism and exhibitionism (peeping Toms and flashers) are the best-known paraphilias because they're common characters in comedy routines. But paraphilias are no joke. When these rituals are directed at nonconsenting participants, they are crimes with serious consequences, and anyone who suffers from a paraphilia is caught in an unbearable bind. For them the only escape from shame and self-hate is to redirect these feelings into episodes of white-hot intensity, often lasting for hours, and frequently with multiple orgasms. Afterward any satiation is remarkably short-lived, and before long the same torturous sequence demands repetition. When they try to fight their urges, which they inevitably will, the intensity only grows. No erotic predicament is more exciting—or painful.[6]

Carlos: Thief of thrills

By the time I met Carlos he was at his wits' end. For the second time in as many months, college classmates had complained to the authorities about his loitering at the gym, spying on them as they undressed and showered. More than once he had been observed furtively masturbating. Several of the guys were getting very angry, some even threatening violence. With his college career in jeopardy and under the threat of legal sanctions, he was awash in shame and self-reproach.

Carlos had struggled with negative beliefs about himself for as long as he could remember. Ironically, prior to the brouhaha at the gym, he had been feeling much better about himself. In fact, what upset him most—other than the humiliation of being "found out as a pervert"—was that his growing self-esteem, painstakingly cultivated with the help of a previous therapist, seemed suddenly to have vanished. Carlos was aware that his sexual troubles were linked to his unhappy childhood, although he wasn't sure exactly how. As his story unfolded, the links grew clearer.

His family emigrated from Central America to a small town in the southwestern United States when he was an infant. At home a chronically negative, embittered mother invariably focused on what was wrong with everyone, expressing her disapproval with hostile sarcasm. His father, who struggled to eke out a living at low-paying jobs, was a prime target of his wife's demeaning jabs. Carlos remembered him as stoic, silent, usually absent, and unwilling or unable to counter his wife's attacks. Carlos and his two siblings were left to fend for themselves.

Things weren't much better in the neighborhood. The white majority, mostly poor themselves, vented their frustrations on anyone different, especially Hispanic immigrants. Carlos regularly retreated into a world of comic book heroes and dreams of being a celebrity. When he discovered masturbation at age twelve, his fantasies and the orgasms they produced forced him to confront what he had always tried to ignore—he was undeniably attracted to other males.

From afar he admired the popular guys, especially the handsome jocks with muscular builds, the ones who exuded confidence and bravado. His admiration for them increased in direct proportion to his self-hatred. Placing them high on a pedestal, he gazed at them with a mixture of awe, envy, desire, and resentment—for being so much better than he, and sometimes for teasing him ruthlessly. His fondest, seemingly impossible, dream was that one of them would befriend him and initiate sex. His eroticism was taking shape with him cast as the unworthy outsider. When he was fifteen a pivotal event locked him into that role.

One day after school Carlos was the last one in the shower after a long swim. Suddenly, in walked Drew, one of the guys he admired most, a star of the swimming team who had rarely so much as acknowledged Carlos's existence. This time he greeted Carlos with a smile and bantered with him playfully in the shower. As they talked, Carlos was transfixed by the strength and beauty of Drew's body. He tried to act casual in spite of having a full erection, a fact that Drew commented on with none of the derision Carlos expected. Their conversation turned to girls, to sex, and to feeling horny.

As they dried off, Drew motioned toward the towel room door. Inside he reclined on a pile of damp towels, his penis now fully erect. He let Carlos rub and lick him everywhere. Reveling in the taste and feel of Drew's genitals, Carlos brought him to orgasm while masturbating himself. "You're a great cock sucker," said Drew, "do it again"—an order Carlos eagerly obeyed. Afterward they dressed silently and left. Except for two other chance encounters, Drew continued to ignore Carlos most of the time. Carlos spent countless hours, however, waiting for opportunities to be alone with Drew. With great enthusiasm Carlos told me, "I'd never felt so alive, so *accepted.*"

The next year Carlos and his family moved to California. As a kid he had concluded, with the encouragement of his negative mother and racist community, that Hispanics were inferior. Now he saw plenty who were thriving, some of whom became role models for the success and self-respect he craved. Similarly, he met gays who celebrated their sexuality and formed the intimate, romantic bonds that he had always assumed were available only to straights. Gingerly, he began to come out of the closet.

Paradoxically, as his self-esteem increased, his voyeurism became more obsessive. Whereas he used to cower in the shadows, now he became bolder, taking more risks, almost demanding to be noticed. And so when the complaints came, he was certainly devastated but also gratified to have finally become sufficiently visible to upset someone—especially the men to whom he mercilessly compared himself.

Compulsive voyeurism, like other paraphilias, is relatively rare. Yet it's important to keep in mind that millions of men and women are regularly stimulated in nonobsessive ways by the very things that excited Carlos. Who hasn't been titillated by catching a glimpse of someone undressing or overhearing a sexual discussion or encounter? Most of us are also familiar with the bittersweet thrill of feeling inferior to those who most strongly attract us. And men and women of all sexual orientations, especially males in their teens and twenties, can identify with Carlos's sexual preoccupation and constant quest for visual stimulation.

Carlos's eroticism was problematic because his negative core beliefs required that he stay in an inferior position. Yet within the self-defeating framework of his CET, Carlos used virtually every known source of arousal, including all four cornerstones of eroticism. The chasm between Carlos and the men he worshipped unleashed a flood of yearning. At the same time, a furtive sense of naughtiness permeated every scene, highlighted by the ever-present risk of discovery and punishment.

The entire drama was further energized by a push-pull dance of power. On one hand, Carlos was clearly submissive to the men whose very existence seemed to mock him. Yet following the lead of his masturbation fantasies, primordial images of aggression and conquest helped him turn the tables. If the attention and reciprocation he craved weren't freely given he would steal them with stealth and cunning. He stalked the men he envied as prey, using them without their consent as pawns in his psychodrama. Finally, a forceful undercurrent of ambivalence toward everyone—himself as well as the men who simultaneously excited and demeaned him—added yet another dimension to an already explosive concoction.

The entire scene was infused with plentiful and intense emotions. Some were positive, such as the genuine admiration and appreciation he felt toward the men who represented his ideals of masculinity. Negative emotions included resentment, hostility, fear, guilt, and shame. As Carlos explored his eroticism more deeply, he discovered that revenge was a particularly gratifying aphrodisiac. It was both frightening and exciting to be spotted by the men he stalked. Only if they knew what he was doing could they be made to squirm, as other men in the past had made him squirm. He savored the notion that they felt humiliated when Carlos used them as pawns in his sexual games.

One thing Carlos's eroticism did *not* allow was the reciprocation of love and affection. He was trapped in the same bind as everyone whose eroticism is built on a foundation of self-hate: anybody who might be attracted to him was, by definition, excluded from the ranks of the desirable. Only those who reinforced his self-contempt were worthy objects of desire.

Gradually, Carlos discovered why increasing self-esteem was making his voyeurism even more obsessive and problematic. His movement toward self-acceptance was threatening to unravel the central purpose of his eroticism: to turn the trauma of self-hate into the triumph of sexual excitement. He was torn about how to respond.

EROS IN CRISIS

Anyone whose eroticism is founded on antiself core beliefs and who later develops more self-loving attitudes and behaviors will eventually face an erotic crisis similar to Carlos's. *Webster's* defines "crisis" as "a time of great danger or trouble, whose outcome decides whether possible bad consequences will follow."[7] An erotic crisis is launched by the realization that the requirements of self-esteem and the lure of firmly established sexual patterns have become incompatible.

For some this realization is gradual, beginning as a vague hunch that cherished turn-ons are somehow working against them, a hunch that builds until it demands attention. Often an erotic crisis grows out of a crisis of another kind, as it did for Regina when she cut herself, or for Carlos when he was threatened with expulsion for his voyeuristic activities. No matter how it begins, every erotic crisis ultimately reaches a point at which painful truths must be confronted and life-changing decisions contemplated. No one chooses such a crisis. Many will do almost anything to avoid it. And rarely, if ever, is any obvious solution in sight. It is indeed a time of great trouble.

HIDDEN MANIFESTATIONS OF SELF-HATE

Ironically, an erotic crisis is often a sign that significant growth is under way. Some people unexpectedly stumble into erotic crises while working on problems that seem far removed from sex. Perhaps they are making strides toward greater confidence and self-affirmation when, much to their consternation, their progress inexplicably stalls. More than a few of these men and women come to realize that feeling better about themselves—even in nonsexual ways—has unexpected erotic consequences.

Nick: Unacceptable choices

Nick entered therapy at the urging of his boss. Though he had once been an up-and-coming advertising executive, for at least two years his career had gone nowhere. In recent performance evaluations he had been criticized for a lack of forcefulness and self-confidence. "I seem to be going backward," he complained. "Maybe I'm out of my league." His current job—managing major marketing campaigns for large corporations—required assertiveness and determination, two qualities he seemed unable or unwilling to muster with the necessary consistency.

In spite of a sharp intellect and an engaging personality, he had always felt intimidated by aggressive men, exactly the type he was surrounded by now. He was certain this had a lot to do with his father, an ex–navy man who placed an exceedingly high value on discipline, physical prowess, and winning at all costs. Nick and his two brothers had been in constant competition in school, sports, and virtually every other activity. Not only did his father praise whoever was on top (usually one of Nick's brothers), he was also quick to belittle whoever lagged behind (usually Nick).

Of the three boys, Nick had the closest relationship with his mother. He shared her passion for literature and music, which his father treated with indifference if not disdain. Also like his mother, Nick responded emotionally rather than with the cool stoicism and logic valued by his father. Nick knew he had an inferiority complex and had obviously put a lot of effort into overcoming it. In the supercompetitive atmosphere of his job, however, he could only see himself as he assumed he appeared to his father—a loser.

As a result of our discussions, Nick decided to practice assertiveness experiments at the office and carefully observe his own and other people's reactions. He made a point of noticing and challenging self-deprecating thoughts and, whenever possible, honoring rather than downplaying his achievements. His efforts appeared to pay off as he began to allow his self-image to depart significantly from the negative core beliefs he internalized long ago. His boss was delighted with the obvious change.

Before long, however, Nick's push toward self-acceptance began to run out of steam. After weeks of steady progress, his enthusiasm for experimentation and growth all but collapsed. He missed therapy sessions and was clearly on the verge of quitting. He looked puzzled and irritated when I explained how common it is for people to resist the very changes they yearn for because they are unconsciously reluctant to defy a disapproving parent. Nick was only vaguely interested.

An intuition flashed through my mind completely out of the blue, and I veered off in a new direction. "Sometimes," I said, surprising Nick and myself, "it's the sexual implications that frighten people." He responded with obvious agitation, "*What* sexual implications?" I proposed that becoming more confident might be having subtle unwanted effects on his sexuality. "Are you saying I need to feel like a wimp to get off?" he asked incredulously. I agreed that something like that was possible but suggested that it might be more a matter of old feelings of unworthiness or inadequacy having become interwoven into his arousal patterns.

Perhaps as much to prove me wrong as anything else (he was much more competitive than he realized), Nick engaged anew in therapy. We turned our attention to his sexual life, which we had scarcely touched on before.[8] All I knew was that he had been seeing Barbara for about a year and that he felt pretty good about their relationship. But I had no idea what attracted him to Barbara, or anyone else for that matter.

Nick revealed the details of his favorite turn-ons slowly and cautiously. With impressive honesty and persistence, Nick discovered that the dramatic themes that stirred his passions did indeed contain strong, yet carefully concealed, links to his boyhood shame about not measuring up.

During adolescence his brothers dated the popular girls, the ones he could only fantasize about. And fantasize he did, developing a favorite image in which several gorgeous women compete for his attention. As they stimulate him in every imaginable way, they become out of control, voracious sex fiends. At the climax of these fantasies he has intercourse with them one by one, while the others caress and kiss him feverishly. As you can see, there's no obvious sign of inferiority in this fantasy. But displays of confidence and bravado are often exciting precisely because they temporarily reverse the feelings of inferiority from which they spring. Nick made a crucial observation: "I've been noticing how this fantasy is most intense when I feel incompetent or defeated. I guess it's how I make myself feel better."

"What turns you on when you're feeling strong?" I asked. After a long silence Nick answered, "I'm not sure. I definitely enjoy sex with Barb when I'm feeling good about myself. But I guess I would call that loving sex, not nearly as torrid as my fantasies."

For millions of men and women, the roles they assume in their external lives bear little relation to their sexual preferences. For example, I've worked with high-powered executives, both male and female, whose fantasies revolve around total surrender to a dominant other. Many have little difficulty accepting the apparent contradiction. But others require a reasonably close match between their public personas and their private turn-ons. Like Nick, they must actually *feel* inferior and unworthy for their CETs to work properly. These men and women have no idea why they hold themselves back or repeatedly place themselves in no-win situations. Some don't even realize what they're doing because the entire process functions automatically—until they make a concerted effort to become conscious of it.

As irrational as it sounds, Nick had placed strict yet invisible limits on his self-esteem to protect the inferior position his CET was designed to fix. He was sacrificing his personal and professional advancement because feeling stronger and more confident was incompatible with his eroticism. Once he saw what he was doing, Nick could no longer act unconsciously. Now he had a choice—albeit an unappealing one. Either he would honor the limits imposed by his negative self-image or he would risk reducing his passion by pushing ahead with his growth.

It may be difficult to fathom how success in one area of life can cause trouble in another. Yet Nick's predicament is not as unusual as it seems. In fact, whenever clients struggle with fears of success I now always suggest they consider the erotic implications of their goals. Once negative core beliefs have been eroticized, most people instinctively attempt to preserve them no matter what the cost.

POSITIVE AVERSIONS

Many clients enter therapy already in an erotic crisis, though they don't fully realize it. At first they simply want to resolve a sexual dysfunction or rekindle waning desire. Only later do they discover a more serious problem: negative core beliefs are wrecking their eroticism. In these cases unpleasant symptoms aren't just necessary, they're ultimately beneficial. They force a person to stop, take notice, and initiate a change.

Brenda: Self-esteem and the death of desire

"My husband, Ernie, sent me here because I'm a mess sexually," Brenda announced with a mixture of pain and embarrassment she tried unsuccessfully to hide. "He *sent* you?" I asked, echoing the implication that she had been towed into the shop for repairs. She explained that she was rarely orgasmic with Ernie but insisted she didn't mind. Until recently, sex had been tolerable, sometimes pretty good. But now Ernie was displeased because Brenda had lost all interest. She couldn't even acquiesce passively as she had so often done before. "I don't know why I'm so screwed up," she said, sighing, "but the thought of sex almost makes me sick.

"Pretty crazy, huh?" she asked, watching closely to see how quickly I would agree with her self-criticism. Her eyes moistened when I insisted that there must be very good reasons for her to feel so disgusted with sex. When I asked if Ernie had considered coming in with her, she replied incredulously, "Are you kidding? Ernie would *never* go to therapy!"

Although their marriage was clearly in serious trouble, intuition told me that Brenda had come for more important reasons. Her voice mostly communicated sadness and resignation, but occasionally she would slip in little clues that she was on the threshold of a radical new experiment: to find out what *she* felt and wanted. Wearily, she accepted my invitation to explore her sexuality for her own purposes rather than Ernie's, to see if we could uncover the messages her inhibitions were trying to convey.

Born into a large, prominent New England family, Brenda had been terrorized by a loud and pompous father who relied on intimidation and violent flare-ups to get his way and keep everyone around him in a state of fear. Like her mother, Brenda adopted a submissive "low profile" in hopes of avoiding her father's wrath. She excelled in school and always had close girlfriends, but males seemed too frighteningly crass and aggressive. Yet in one of those predictable ironies of erotic life, her gentle ways attracted men with an urge to dominate. Although reluctant to admit it, she finally acknowledged that these attractions were quite often mutual. One day she confided in a hushed, confessional tone that in her favorite fantasy an almost savage stranger sweeps her off her feet, carries her to bed, and ravages her, catering to her every need—a common fantasy among women, and quite a few men too.

In each of her three significant romantic relationships she adopted the compliant role she had learned so well, yielding to almost every demand, sexual or otherwise. At the beginning of her relationships, when her lovers were on their good behavior, she enjoyed sex. Gradually, however, her enthusiasm would decline as she surrendered herself to the supposedly more important needs of yet another angry man.

She and Ernie met as she was beginning graduate studies toward a degree in special education. He was an ambitious law student to whom she was instantly attracted. It was his idea to get married. Even though she had always wanted to be a teacher, Ernie insisted that she drop out of school so she could work more hours and "be there for him." Many years later, when he became a partner in his law firm, she stopped working and tried to get pregnant but couldn't.

Ernie had turned out to be as controlling and unpredictably violent as her father, but she claimed to be devoted to him. "At least he doesn't *hit* me," she once explained in his defense, moments after expressing just a hint of the resentment that seethed behind her soft-spoken facade. When he blew up he would berate her with venomous insults. Then he would be remorseful for a day or two and become the attentive, passionate lover of her fantasies. Only then would she become fully aroused and sometimes orgasmic. After eight years of these predictable cycles, she felt as if life was passing her by. Yet when friends urged her to leave him, a sense of resignation descended upon her and she would remain immobilized.

Something inside was beginning to stir. About two months into therapy she was haunted by a vivid, unsettling dream. In it she's a young girl in the passenger seat of a car speeding down a winding road in the dead of night. Behind the wheel is a "dark man" who seems like her father, although she can't see his face. She wants to steer the car and grabs the wheel. He brushes her away. Frustrated, she tries again, this time more determined because she's "feeling her oats." When she fails again she protests. Then the dark man whirls around, towering over her menacingly, revealing a monstrous face with piercing, enraged eyes. As he slaps her down he thunders, "No!" She awakes in a cold sweat, shaken by the dream's unambiguous message: violent, overpowering men will not allow her to steer her life. She's small and helpless—but getting angry.

As we discussed her dream I suggested that not only was she married to one of these men, but the monster who slapped her down was also within her. *She* wouldn't allow herself to take the wheel. *She* terrorized herself. At first she resisted this idea vigorously, with a hint of exhilaration in her voice. She felt that I was blaming her for her troubles and was defending herself for a change. She was genuinely shocked when I told her how good it felt to see the strong Brenda.

Immediately, she slipped back into familiar territory and began berating herself. And then, in one of those moments that remind me why I became a therapist, I watched as she interrupted herself midstream. "I've got to stop doing *that,* don't I?" We nodded and smiled in unison, both knowing she had reached a point of no return.

Brenda's loss of desire was a subconsciously directed act of self-respect. Her body was expressing what she could not say in words: I refuse to be turned on by someone who treats me as if I'm worthless. Loss of desire is among the most common of all sexual problems. Sometimes this is an inevitable side effect of today's stress-filled lives. Sometimes it springs from unresolved resentments and conflicts in long-term relationships. Sometimes the culprits are the predictable routines that squeeze out the last vestiges of surprise.

More often than you might think, however, diminishing sexual desire is a sign of growth, as it was for Brenda. In such cases lost desire is not a problem to be fixed but a message to be heeded. If a person's CET consistently places him or her in a demeaning position, increasing self-esteem will usually have an unmistakable antiaphrodisiac effect. It is a signal that "business as usual" is no longer possible and a call for fundamental change.

Almost one year after Brenda and I began our work she confronted Ernie with her dissatisfaction and insisted that he accompany her to a marriage counselor. True to form he refused, ranting at her about how sick and tired he was of having a frigid, complaining, good-for-nothing wife. Within two weeks Brenda moved in with a friend and filed for divorce. Although for some time to come she grieved about all she had lost and all she had endured, even during her worst days she never doubted the rightness of her decision.

Away from Ernie, Brenda began to get promising clues about how her eroticism might be different without self-deprecation and fear as its guiding principles. About one thing she was absolutely certain: she would *never* again allow anyone to mistreat her. It was her turn to take the wheel.

CONFRONTING SELF-HATE

In the same way that the troublesome turn-ons in the last chapter required conscious recognition for healing to begin, low self-esteem cannot be resolved until it is confronted. Most people who have turned against themselves recognize how unhappy they are and want to change. I've also observed that people with very low self-esteem usually have pretty clear ideas about how they got that way. The most difficult task is identifying the insidious effects low self-esteem has on every part of their lives.

If you're aware that you generally perceive yourself in a negative light, chances are those feelings are expressed in your eroticism too. The stories in this chapter illustrate four primary ways that negative core beliefs can distort erotic experiences: (1) by disrupting your ability to accept sexual pleasure, (2) by creating an overwhelmingly intense, compulsive urge to act out a ritualized scenario, (3) by drawing you into self-defeating situations, and (4) by restraining you from developing your full potential—not just sexually but in other areas as well.

Fortunately, you've already learned a great deal about the storylines that animate your most exciting turn-ons. Embedded in your CET are habits of self-perception you use to position yourself in relation to objects of your desire. By focusing your attention on the recurring themes of your eroticism you can develop a clearer picture of how you evaluate your worth before, during, and after exciting fantasies and real-life encounters.

One valuable exercise is to recall as vividly as possible how you felt about yourself when you first became aware of sexual curiosity or desire. If you considered yourself ugly, unworthy of love, or destined to be despised and rejected, chances are these beliefs—and your attempts to counteract them—were part of your erotic development. You probably have clearer memories of sexual fantasies and experimentation following puberty. Can you identify any early experiences in which self-devaluation and arousal were somehow linked together? By examining the roots of your eroticism with an eye toward your attitude about yourself, you can gather valuable clues about the core beliefs—both negative and positive—that have shaped the person you are today.

As you've seen in this chapter, self-hating core beliefs, especially the ones that result from having been treated with contempt or indifference, are extremely deep-rooted and resistant to change. However, the human spirit is incredibly resilient. Through self-discovery and nurturance, the tiny voice of self-affirmation grows stronger. Listen closely and it will guide you out of self-hatred and despair. With patience and compassion the wounded self will heal and thrive.

8

WINDS OF CHANGE

Seven steps point the way to sexual healing and growth.

Now that you've looked at some of the most perplexing, persistent, and unpleasant dilemmas of erotic life, you may be wondering how they can ever be resolved. If you or someone you care about has struggled with a troublesome turn-on, you know firsthand that it's useless merely to wish for change. What's needed is a plan of action that mobilizes your innate capacities for healing and renewal. This chapter is devoted to such a plan.

It is important to realize that change is intrinsic to the erotic adventure. For instance, as you discover new avenues to pleasure, your sexual repertoire naturally expands to accommodate them. Unfortunately, we must confront an unwelcome irony: the erotic patterns most in need of modification—the ones that create turmoil as well as excitation—are typically the most resistant to change. In other words, the more troublesome a turn-on, the tighter its grip.

Common sense might lead you to expect that the pain of problematic sex would be a powerful motivator for change. And in a sense this is true. People whose sexuality has become a source of distress usually want to change. Yet the same conflicts the person hopes to resolve may also produce extraordinarily high levels of intensity. Consequently, those who suffer the most from their erotic patterns often feel the most compelled to repeat them.

As a result of my work with hundreds of men and women who have successfully modified or expanded their turn-ons, I've identified seven pivotal steps that consistently lead to positive erotic change:

Step 1. Clarify your goals and motivations.
Step 2. Cultivate self-affirmation.
Step 3. Navigate the gray zone.
Step 4. Acknowledge and mourn your losses.
Step 5. Come to your senses.
Step 6. Risk the unfamiliar.
Step 7. Integrate your discoveries.

Steps 1 and 2 set the stage for change by helping you define what you're trying to accomplish and how you expect to benefit. Steps 3 and 4 prepare you for two difficult challenges: feeling lost without a map as you leave the comfort of the familiar and coping with the grief that is a commonly ignored aspect of significant change. Steps 5 and 6 show you how to experiment with specific behaviors and use them as guides to your untapped potentials. Step 7 is concerned with making change endure.

STEP 1:

CLARIFY YOUR GOALS AND MOTIVATIONS

Some of the most welcome changes occur spontaneously. Eroticism is constantly evolving as old turn-ons no longer satisfy and new possibilities for enjoyment appear as unexpected gifts. Under the best of conditions these changes are propelled by pleasure, without any need for conscious goal-setting.

Unfortunately, not all sexual changes are so easy or pleasurable. Well-established erotic patterns often continue exerting a major influence long after they have ceased to be truly satisfying. Sometimes we lose touch with the pleasure-based motivations that normally guide us. When desired or necessary changes aren't happening automatically, we need to think long and hard about exactly where we wish to go. Then, by clarifying our reasons for wanting change, we can unleash the necessary energy and determination.

THREE KEYS TO GOAL-SETTING

The kinds of goals you set and how you go about setting them can make the difference between moving forward and remaining stuck. In my experience there are three crucial guidelines for successful goal-setting:

1. Keep your goals simple, specific, and limited. Vague goals such as "I want a better sex life" or "I want more fulfilling relationships" are virtually useless. It's not easy to define the essence of what you hope to accomplish, but the rewards are well worth the effort. "I want to be able to enjoy sex with someone I love" or "I want to learn how to get turned on without feeling anxious," are examples of much more useful goals because they point you in a specific direction. Whenever you consider a goal ask yourself, "How will I know when I get there?" If you can't say, then your goal probably isn't sufficiently clear.

Try to avoid goals that call for sweeping change. Many people erroneously believe that building up a case against their entire eroticism will somehow increase their motivation for change. But if you believe you have to change everything you will end up changing nothing. Whenever I hear a client say, "My sexuality is completely screwed up," I know his or her immediate goal is self-criticism—not change.

2. Pay much more attention to what you want than to what you don't want. One of the biggest mistakes people make when they contemplate erotic change is to focus on what they don't want to do—don't be so inhibited, don't be self-conscious, don't ejaculate so fast. When erotic patterns are causing distress it's easy to see how getting rid of problematic behaviors can look like the obvious solution. Those who think of themselves as sex addicts are especially prone to this mistake. They don't realize that *struggle is the fuel for all compulsions.* When they fight with themselves they become more compulsive and out of control, not less. If you're serious about changing, express your goals in the affirmative, with the emphasis on your ultimate destination and the concrete steps necessary to get you there.

3. Honor your individuality. Helpful goals must be specifically tailored to who *you* are. The key is to begin envisioning your eroticism in its healthiest form. It's a waste of time to base goals on abstract concepts about how eroticism "should" operate. Beware especially of goals that call for fundamental changes in your personality, for they will ultimately undermine your self-esteem and leave you feeling discouraged and helpless.

Be cautious about setting goals that lie completely beyond the scope of your experience. Sometimes this is necessary, of course, as in the case of someone who wants to feel sexy toward a lover or spouse even though he or she has never experienced arousal in conjunction with affection. Whenever possible, it's a good idea to link your goals to experiences you've actually had in the past that are close to ones you'd like to have more often. With rare exceptions, even people whose erotic patterns are profoundly problematic can, if they are willing to look closely enough, identify moments that offer glimpses of their potential.

Goals that involve altering your sexual attractions are especially tricky. Attractions are as individual as fingerprints—and almost as difficult to change. Yet I've known plenty of men and women who were consistently attracted to a particular "type," only to find they have fallen head over heels for someone they would never have considered desirable before. Nevertheless, keep in mind that attractions can rarely, if ever, be changed through willpower.

If you want to select different kinds of partners, be as clear as possible about which specific aspects of your attractions have caused you trouble. Most people who attempt to change their attractions make the mistake of assuming that their preferences need to be entirely transformed. This is rarely necessary. As we shall see, positive change frequently involves using old patterns in new ways.

TWO MOTIVATIONS: PUSHES AND PULLS

Like beacons, goals give us direction. But to mobilize the energy to move us toward these goals we need motivation. There are two fundamental types of motivations: pushes and pulls. It is important to distinguish between them.

Push motivations prod you into action when the discomfort of the status quo becomes intolerable. Unpleasant emotions—especially fear, guilt, hurt, loneliness, and desperation—are examples of compelling push motivators. For instance, Maggie wanted to break her penchant for chasing unavailable men because constant feelings of rejection were becoming unbearable. She was also afraid that time was passing and she might never find a lasting relationship.

Similarly, even though Ryan was tremendously excited by his compulsive pursuit of forbidden thrills through phone sex and porn, he felt increasingly distraught about his lack of choice in the matter. He was also haunted by images of being found out, ridiculed, and cast aside by family and friends. And he didn't like being more interested in raunchy sex with sluts than loving sex with his girlfriend.

Different people are able or willing to tolerate different levels of distress before they actively pursue a change. If someone's behavior causes them *only* distress they will probably change it quickly and with relative ease—unless an unconscious need for self-punishment compels them toward destructive repetitions. Change is more difficult when troublesome behavior is both distressing *and* rewarding.

Unlike push motivations, which require you to face your distress and be made uncomfortable, *pull motivations* promise potential rewards and benefits. Believable, enticing images of how your life can be more fulfilling are immensely helpful in keeping motivation alive even in the face of adversity.

Maggie was willing to consider stepping outside the familiar framework of unfulfilled longing not simply because she feared a life of loneliness but also because she had experienced moments of actual intimacy and was beginning to believe she deserved more of it. Similarly, Ryan was not only pushed by shame about being a prisoner of his taboo impulses, he was also pulled by a genuine desire to feel closer to his girlfriend.

ANTIMOTIVATIONS

When you set out to change an entrenched pattern, sooner or later you will feel inclined to back away from the very changes you seek. Psychologists call this resistance. All kinds of doubts and fears, both conscious and unconscious, can contribute to resistance. What if you fail? What are the implications of success? Could tampering with your eroticism ruin it? Will friends, family members, or sexual partners encourage you, or might they be threatened by what you're doing? If your sexuality changes will you still recognize yourself? These concerns, along with a host of others, are *antimotivators* because they prevent you from vigorously pursuing your goals.

Antimotivators are as much a part of the change process as your goals and motivations. Because antimotivators are most likely to undermine change when they operate subconsciously, it's smart to become as aware of them as possible. If you face your fears about change rather than suppress them you will be much better able to address what concerns you. Antimotivators contain crucial information about which courageous acts are called for—and when.

YOUR MAP FOR CHANGE

Successful and fulfilling erotic changes involve all three: push motivators, pull motivators, and antimotivators. To boost your awareness, put them all together on one piece of paper so you can see them at a single glance. Write the headline GOALS, and then list your objectives as clearly as you can.

Maggie's Motivational Map

Goals:

- To be attracted to someone available for a relationship
- To *only* love someone who loves me in return
- To develop other turn-ons besides longing
- To let myself accept love

Push Motivators (why I need to change)	**Pull Motivators** (potential benefits and rewards)	**Antimotivators** (fears/resistance to change)
• Hate feeling rejected • Afraid I'll end up alone • Sick of feeling depressed, worthless, and unlovable • Wasting too much time waiting around for the phone to ring	• Feeling the joy of intimacy and commitment • Relaxing, not always having to chase or worry • Having fun when *I* want to • Sharing special occasions with my boyfriend • One day getting married and having a family	• Hard to believe someone desirable could want me • Afraid I can't feel passion without longing • Not sure available men will attract me • Afraid I'll be hurting my mother if I find love

Next divide the area below your goals into three columns. Label the left column *Push Motivators.* As specifically as possible, list the reasons that you believe change is necessary. Focus on what concerns you about your eroticism—anything that causes pain or distress. Label the middle column *Pull Motivators.* Here spell out how you hope to benefit, either immediately or in the long run. Label the right column *Antimotivators* and list any fears and hesitations that might hold you back.

Resist the urge to edit as you write. Grant yourself freedom to be as irrational as you want. This map is for your eyes only, so no one else is going to judge it. Once you've listed everything you can think of, place a star next to the items about which you currently feel most strongly.

Now you have a picture of where you're trying to go, why you want to get there, and the roadblocks you're likely to encounter along the way. Keep in mind that goals and motivations are not static. Expect them to evolve as you become increasingly experimental. One of the advantages of writing all this down is that you can return to it, refining your list as you move along. See Maggie's motivational map on the previous page. Notice how she has succinctly summarized all the key elements for her erotic growth.

STEP 2:

CULTIVATE SELF-AFFIRMATION

Those who set out to alter a troublesome turn-on commonly begin in a self-critical posture, vigorously focusing on what's wrong with them. They doggedly cling to the conviction that somehow their judgments will bring about change. This mistaken belief is a major impediment to growth.

I'm not denying that constructive self-criticism can be a valuable stimulus for action. But when self-demeaning thoughts and feelings predominate, they ultimately sabotage your innate capacities for healing. The only alternative is to cultivate the ways of self-affirmation. To affirm yourself is, first and foremost, to assert your right to take up space in the world. Beyond that, it is an ongoing process of developing and expressing more of your truest self.

Many confuse self-affirmation with narcissism, a preoccupation with oneself that makes it difficult to recognize anyone else's feelings and needs. In fact, narcissistic adults who seek continual recognition and praise are trying to compensate for a lack of genuine self-affirmation.

FROM STRUGGLE TO CONSENSUS

The moment you decide to make a change you initiate a series of interactions among two or more internal aspects of yourself. Perhaps one part of you has reached the end of its rope and is now desperate for change no matter how difficult, while another part clings to familiar patterns in spite of the pain and dissatisfaction they cause. Yet another part may only be concerned that you conform to external standards of behavior.

Ever since I successfully quit smoking after countless failed attempts, I've been fascinated by how people prepare themselves for change. Everyone I talk to who's made a difficult transition instantly understands what I mean when I say that I was able to quit smoking because my entire being eventually came to an *inner consensus*. But I have yet to meet anyone who can explain exactly what this consensus is or how it is reached. Of two things I am sure. First, within you is a unifying force known as the Self that holds the key to reducing destructive internal conflicts.[1] Second, compassion is the master emotion of positive change.

MOBILIZING THE POWER OF COMPASSION

Compassion is one of the most healing of all responses to pain and suffering—for both the giver and the recipient. It is the opposite of malice or indifference. When you feel compassionate toward someone who is hurting, you feel no inclination to judge them. In the same way, whenever you are able to direct compassion inward, self-criticism temporarily dissolves.[2] Especially for those who grew up in abusive environments, healing depends on compassion.

Those who are not intimately acquainted with compassion typically equate it with self-pity—which it is not. Compassion does not pity, it *understands.* Sometimes self-pity is a perfectly valid response—to a painful loss, for example. But most people recognize that chronic self-pity depletes their vitality. In stark contrast, compassion mobilizes energy and is a wellspring of courage.

Both self-affirmation and compassion are built up when you make both large and small decisions that promote your best interests. An effective way to recognize your capacity for positive action is to remember choices you've made in the past that were clearly good for you. Trust the memories that spontaneously come to mind. Perhaps you'll recall committing yourself to a relationship in spite of the risks or leaving one that was undermining you. Other possibilities might include moving to a new location (not because you were running away, but because you always wanted to do it), quitting a dead-end job or embarking on a new career, confronting an addiction or destructive habit, making health-promoting choices about food or exercise, developing a new skill, or doing something that felt right for you despite disparaging comments from family or friends.

Focus on how you reached those decisions. Did you realize at the time that you were acting in your own best interests, or did that become apparent later? Were your choices spontaneous, or did you struggle with them for a time? How did you deal with your misgivings and fears?

As important as these dramatic turning points are, the cultivation of self-affirmation as a way of life depends on the cumulative effect of smaller decisions made on a daily basis. Such choices may seem so insignificant at the time that you hardly notice them. Yet their impact can be enormous. Think back over the last two weeks. How many instances can you recall, no matter how small, in which you acted in ways that were self-validating? Such instances might have involved expressing feelings or opinions that you would normally keep to yourself, assertively asking for what you wanted, saying "no" in spite of pressure to acquiesce, or taking care of yourself in other ways.

You may wonder if being self-affirming on a regular basis

might make you demanding, pompous, blind to your own faults and imperfections, or less sensitive to others' feelings. In my experience, increases in self-affirmation actually have the opposite effect—they make us less self-centered and more cooperative and empathic.

What does self-affirmation have to do with erotic transitions? Just about everything! With few exceptions, blockages and distortions of your deepest inclinations lie at the root of self-defeating turn-ons. The alternative is to listen carefully for the guidance that comes only from within.

STEP 3:

NAVIGATE THE GRAY ZONE

Once a significant transition is under way, it's only a matter of time until you enter a period of awkward uncertainty when you're no longer where you've been, but you haven't arrived at where you're going. Welcome to the *gray zone*, which, of course, is not a location but a state of mind distinguished by a distressing absence of clear pathways and landmarks. For a time you feel as though you're wandering aimlessly, disoriented, lost without a clue about how to regain your bearings.

The more important and challenging the changes you seek, the more prolonged will be your stay in the gray zone. For some, being in the gray zone feels like standing at the edge of an abyss. But if you can tolerate its ambiguities, the gray zone holds unparalleled opportunities for self-discovery. Gradually, you will notice that the gray zone is not at all the featureless desert it first appears to be. It's more like a blank canvas on which you can experiment with new shapes and colors.

STUMBLING INTO THE GRAY ZONE

Sometimes the first hint that you are entering the gray zone is a realization that partners, situations, or fantasies that have reliably turned you on in the past are losing their allure. If you aren't prepared for the surprises that await you in the gray zone you might misinterpret your waning interest as a sign of trouble rather than a harbinger of positive change. Men find it especially difficult to handle this temporary reduction in desire because it is usually reflected in less reliable, softer, or nonexistent erections.

When old turn-ons begin to lose their effectiveness, some people embark on a search for more intense forms of stimulation to prove to themselves that everything still works. Unfortunately, their first impulse is often to repeat—with even more single-minded determination—the very patterns that are losing their grip. One man could no longer ignore the fact that he wasn't responding to his porn collection featuring leather-clad dominatrixes. So he went on a frantic search for new porn with even more exaggerated images of dominance and submission. Only later did he realize that these purchases were completely useless because his eroticism was evolving away from the imagery of power and toward—well, he didn't know yet. But once he accepted the fact that his old arousal patterns were crumbling to make room for something new, yet undefined, he was better able to accept the flux and uncertainty of the gray zone.

Maggie Revisited: Turning away from longing

When Maggie entered therapy to deal with the loss of her relationship with a married man, she had no idea her eroticism was on the brink of a radical shift. She fully expected her attractions to remain essentially the same. She just wanted to make them work better. But once she realized that she was "hooked on longing," that her greatest turn-ons involved "*almost* being loved," and that fantasies of fulfillment excited her more than the real thing, she was understandably perplexed.

"What do you expect me to do now?" she demanded in exasperation. "I don't even know which men turn me on anymore. I thought therapy was supposed to make me *less* confused. I've never been so unsure!" Maggie had stumbled into the gray zone and didn't want to be there. But out of necessity she rallied her considerable inner resources to meet its challenges.

She was amazed by a curious phenomenon she named the

"dual response." "When I see the kind of man who has always attracted me," she explained, "immediately I'm as fascinated as ever. I can spot these guys across a room. But now the moment they grab my attention, something inside me snaps and I feel myself turn off and I think, 'Yuck! What a jerk!'" This contradictory response is typical in the gray zone. Maggie ultimately discovered that by selecting ambivalent men to pursue she had distracted herself from the fact that *she* was as ambivalent about intimacy as her lovers. "At least I used to enjoy the chase," she said. "Now all I see is the same old story with the same shitty ending: poor little Maggie chasing after crumbs. I've had it! It's too damn humiliating."

Maggie's most important decision was to solidify a policy of self-respect. Her manifesto became: *I don't want to be with anybody who doesn't want to be with me.* "I don't care if I *ever* fall in love again," she insisted. "If he doesn't love me back, I'm not interested." For months she dated only sporadically, gingerly sampling different kinds of men and conducting experiments that had been impossible when she was focused on pursuit. Through it all her policy held firm. She had broken the spell of the chase.

CHOOSING A MORATORIUM

There are times when it is advantageous to *create* a gray zone consciously and deliberately by voluntarily abstaining from your usual practices for a period of time—sometimes specified, sometimes not. Freely choosing not to do what you normally do can help you in two ways. First, by stepping back from automatic behaviors you gain a fresh perspective on what your eroticism is trying to accomplish or express. Second, by detaching from well-worn habits you open up space in your erotic landscape for experimentation and self-observation.

Unless it is imperative that you stop what you're doing right away because of life-threatening consequences (risking arrest, injury to another person, or exposure to AIDS, for example), it's pointless to attempt such a moratorium until an internal consensus is reached. Until then, the most effective approach is to *decide* to continue doing whatever you feel compelled to do, and to do it *mindfully.* Sometimes even a modest increase in consciousness brings people face to face with how their behavior actually makes them feel. For instance, those who are sexually driven often become so "spaced out" that they have little or no awareness of what they're feeling before and during a compulsive episode. Consciousness typically reveals a host of unpleasant emotions wrapped up in the turn-on, most commonly fear, hatred, sadness, and shame. As these unpleasant feelings are recognized, they serve as push motivators, prompting the person to pause and consider a new path.

TAPPING THE POWER OF IMAGINATION

Your journey through the gray zone will yield maximum rewards if you use your mind's eye to perceive possibilities that don't yet fully exist. This may be difficult if you've learned to trust only what you can verify through reason or the senses. Great discoveries—whether in science, the arts, or personal life—require *imaginative leaps* in which the discoverer moves beyond typical modes of thought and perception.

With a little practice and patience almost anyone can learn to apply the gifts of imagination to erotic transitions. There are two requirements. First, clear some time to imagine. Second, resist the urge to criticize the products of your imagination prematurely. And remember, when judged by normal standards of productivity, imaginative pursuits look suspiciously like play. Those who think they must always be accomplishing something are often reluctant to let their imaginations roam freely.

Try this experiment. When you're not preoccupied or in a hurry (is there such a time?) find a quiet, private place to sit or recline. Allow your breathing to slow as it deepens. With your eyes closed, scan your body from head to toe, allowing each group of muscles to relax. As you inhale, notice sensations of calm and warmth flowing through your nostrils, filling your lungs, and radiating throughout your entire body. Each time you exhale, let tension and worry melt away. Don't be concerned if you're easily distracted. Each time your mind wanders, gently bring it back to the soothing rhythm of your breath.

Now recall an especially positive and fulfilling sexual experience. Maybe you'll revisit one of the peaks you've remembered before, or perhaps a different memory will surface. Remind yourself that the most satisfying experiences are not necessarily the wildest. Savor the feeling of actually being fulfilled. Linger over the simple details that brought you pleasure. Reexperience how you felt about yourself during and afterward. Vividly reliving experiences of sexual satisfaction helps clarify your goals and strengthens your motivations.

You can also use your imagination to visualize what might happen if you were to let your eroticism venture off in a new direction. Don't worry if your images are fuzzy or keep changing. By its very nature the imagination plays with possibilities. If at first you draw a blank when it comes to your erotic future, don't be concerned. In time your imagination will conjure up pleasurable fragments. Be aware, however, that the moment you try to freeze one image to the exclusion of others, you halt the imaginative process.

Long before Maggie adopted her policy of turning away from the sorts of men she once pursued, she visualized the implications of such a decision in her mind. Her initial images were of being stuck with one of the "normal" men who had always bored her. But as she recognized how frustrating it was to be perpetually in a state of longing she grew increasingly fascinated by the image of being loved without reservation, not by a boring man but by an available one. She couldn't help noticing that whenever she had grasped for love that dangled just beyond her reach she had become highly excited but also frantic. But when she pictured herself withdrawing from the chase and letting someone approach her, she felt calm and self-assured.

Whether you stumble unexpectedly into the gray zone or place yourself there by deliberately stepping away from well-worn but unsatisfying behaviors, your journey is destined to be unsettling. Expect to feel lost and alone for a time. Chances are you'll also be frustrated as you grope for answers that rarely come easily. Why put yourself through something so uncomfortable? Besides the fact that you may not have a choice, the gray zone is a powerful impetus for change. It is an opportunity to confront yourself as you never have done before—and to emerge stronger, with a piercing new clarity about what you must do for yourself and how it can be done.

STEP 4:

ACKNOWLEDGE AND MOURN YOUR LOSSES

All growth, no matter how desirable or eagerly sought, involves loss. As outgrown habits fall away, you naturally miss their comfort and familiarity. Every troublesome turn-on has fulfilled crucial functions along the way—or it never would have developed in the first place. Even when it hurts you, it cannot easily be tossed aside.

As the changes you imagine find a place in your life, don't be surprised if emotions associated with grief and loss—sadness, loneliness, emptiness, anger, or guilt—are a part of the experience. Try not to resist, downplay, or deny these feelings; they are essential for healing. Grieving is the way through the pain of loss.

Because thrilling turn-ons are often fueled by conflict and ambivalence, they are frequently the most difficult to change, particularly if painful or traumatic experiences are woven into them. As we become aware of how our turn-ons are linked to unpleasant memories, we can begin exploring more fulfilling, less conflicted erotic styles. Unfortunately, these more comfortable expressions of eros typically involve a distinct reduction in sexual intensity, an intensity that will be sorely missed.

Ryan Revisited: Excitement lost

Ever since his father caught him playing with himself, Ryan had been at war with his sexual urges, a battle that had not only taken an enormous emotional toll but had also provided an inexhaustible source of risqué fascination. When he fell in love with Janet he was forced to face the fact that love and lust had long since become so thoroughly incompatible that he was unable to generate sufficient excitement with Janet to trigger an orgasm.

Ryan's therapy progressed rapidly. He discovered that struggle was the fuel for his compulsive urges and that calling himself a "sex addict" only amplified his conflict. Gradually, because he loved Janet, he freely chose to step away from his "affair" with porn shops and phone sex, concentrating instead on the novel experience of sex with a woman he cared about. He and Janet experimented more freely with sensuality and affection, and both enjoyed a deepening closeness.

In response to this progress Ryan fell into depression. "It makes no sense," he lamented. "Everything is going so well, yet a cloud follows me wherever I go. Everything is flat and colorless." As much as Ryan appreciated his relationship with Janet, he also missed the passion of the raunchy sex he had spent a lifetime both fighting and pursuing, the sex that he had always counted on to distract him from unpleasant emotions.

Although reluctant to admit it, Ryan was in mourning. As it does for so many people in his situation, Ryan's growth stalled until he granted himself permission to recognize the depth of his attachment to forbidden lust. He needed to respect the lost rewards—not just the suffering—of the eroticism he was leaving behind.

THE PAIN OF LOST HOPES

As you've frequently seen, problematic sexual patterns evolve to compensate for unmet needs or to soothe unhealed psychic wounds. One reason troublesome turn-ons are so tenacious is that they express an enduring hope that all can be made well and whole. To some degree, sexual reenactments shield us from the distressing reality that in emotional life what is lost can never be regained.

When you alter an unfulfilling erotic pattern, you might begin to feel, more intensely than ever before, the emotional pain that your CET was originally designed to soothe. Understandably, you might shrink from that sting, choosing instead to endure the dull ache of the status quo. Yet, paradoxically, when you allow yourself to feel your pain fully, you free yourself from it. It is the acceptance, not the denial, of hurt that heals us.

STEP 5:

COME TO YOUR SENSES

Difficult turn-ons come in as many varieties as fulfilling ones. Yet one feature is shared by virtually all problematic erotic scenarios: they ultimately disrupt the relationship between a person and his or her body. The vast majority of troublesome turn-ons are rooted in antisexual messages that contaminate opportunities for children and adolescents to express playful curiosity about their own and others' bodies.

The fact that most people develop ways to express their sexuality in spite of antisexual training is a tribute to the resiliency of the human spirit. Yet many continue to honor early restrictions by subconsciously imposing limits on the amount or types of pleasure they allow themselves to give and receive. One pleasure-limiting strategy—so widespread that it is considered normal—involves focusing erotic attention primarily or exclusively on the genitals and other erogenous zones, relegating the sensuous capacities of the rest of the body to "foreplay."

In the last chapter you saw how the good feelings of arousal can fuse together with painful or traumatic experiences, producing either intractable inhibitions or obsessions. Men and women who were abused as children or assaulted as adults may find that even simple touches are so thoroughly associated with trauma that they can't be enjoyed. Others become trapped in ultra-focused, ritualized patterns of behavior or fantasy or both that are long on intensity but short on pleasure. It's not unusual for those caught up in the pleasure-pain bind, particularly men, to seek out maximum genital stimulation and orgasmic intensity, with little or no interest in sensuality or affection.

One of the most effective ways to facilitate erotic healing is to rediscover the sensuous capacities of your *entire* body. The body has a wisdom that transcends logic. By "coming to your senses" you begin to reconnect with the first source of all positive eroticism.

TOUCHING FOR PLEASURE

One of the best ways to reconnect with your sensuality is to use what sex therapists call sensate focus. Originally designed as a set of experiential exercises to help people with sexual dysfunctions calm debilitating performance pressures, sensate focus is the most enduring contribution of sex therapy pioneers Masters and Johnson. Directing your attention to the concrete world of your senses can be particularly helpful when you're in the gray zone, unsure of what to do next. Whether you're trying to resolve a sexual problem or expand your repertoire for pleasure, you can benefit from experimenting with sensate focus by yourself, with a partner, or both.

I recommend you begin by setting aside time for self-pleasuring. If masturbation is already part of your sexual repertoire, consider expanding your enjoyment by touching your whole body—not just your genitals. It helps to create a conducive atmosphere, perhaps with the help of a hot bath, a crackling fire, or music. You may feel silly at first, but a great deal can be learned by unhurriedly gazing at your naked body in the mirror. Pause to appreciate the parts of your body you like best. But there's no need to pretend if, like everyone else, you also have features you wish were different.

Next sit or recline comfortably and slowly touch yourself everywhere. Notice which places feel good as well as which don't. Adopt a playful, experimental attitude, touching yourself with varying rhythms and pressures. Try using a lotion or oil (safflower, peanut, or coconut oils work well) to add a silky feel. If it interests you, use a vibrator. See if you can discover something new about your body, an unexpected source of pleasure or a way of touching you've never realized could feel so good. If you want to touch your genitals, do it a little differently from the usual way.

Pay special attention to your breathing. Deep, easy breaths allow pleasurable sensations to flow; shallow ones restrict them. If you become aroused—which certainly isn't necessary—notice how your muscles gradually tense. See what happens if you deliberately let your muscles *relax* as arousal builds. Deep breathing and muscle relaxation are the two most direct ways to slow down and stretch out your enjoyment. If you're not used to this leisurely, nongoal-oriented approach to self-pleasuring, you may feel a bit awkward until you try it several times.

Men who want to "last longer," to be able to enjoy more sexual stimulation before ejaculating, can benefit tremendously from experimenting with this slower approach to masturbation. The key is to discover that by taking deep breaths, allowing your pelvic muscles to relax, and temporarily reducing the speed and intensity of the stimulation, you can influence how quickly or slowly your body responds. Keep in mind that these simple techniques are effective only before you reach the point when ejaculation becomes inevitable. Once you learn to prolong your responses in private, you can then use similar techniques with a partner.[3]

Fantasies are often a natural part of self-pleasuring. In fact, during masturbation people typically explore their fantasies more freely than at any other time. Some fantasies contain important clues about what might turn you on with a real-life partner, whereas others are enjoyable only in the realm of imagination and have nothing to do with reality. Either way, see what it's like to let your fantasies roam freely during self-pleasuring.

If you have a tendency to become so caught up in fantasies that they distract you from your body, you can benefit by focusing as much as possible on physical sensations. There are people who are so thoroughly attached to specific fantasies that they believe they can't become aroused or have orgasms without them. For such people sex is a "head trip" rather than a sensuous experience. If this is true for you, it's a losing proposition to fight your fantasies. See what it's like, though, to gently direct your attention to your own touch, even if the intensity is less than you're used to.

Sharing sensate focus with a partner, of course, opens up additional opportunities and challenges. On the plus side, touch usually feels even better when it is provided by another person. And the visual and tactile appreciation of someone attractive brings additional sensuous delights. The difficult part is that the presence of a partner, with his or her own preferences, necessitates some form of communication.

I suggest you begin by taking turns massaging and being massaged. Distinguishing the pleasures of giving and receiving helps to identify how you feel about each experience—which may be quite different. It's important to create an inviting atmosphere just as you did for self-touching. Make a point of spending plenty of relaxed time together before you begin, perhaps sharing an intimate conversation over dinner. Be sure the room is warm. Also, because it can be fatiguing to maintain your balance while massaging someone on the relatively soft surface of even the firmest mattress, it's better to lay a thick blanket on the floor.

Decide who will touch first. Using a massage oil, begin caressing your partner with smooth, slow strokes, ranging from soft to deep. Ask for feedback on which feels best. Encourage your partner to alert you immediately if anything hurts or tickles. Take your time and massage all areas of your partner's body that feel comfortable for both of you. Men and women who have been sexually abused or assaulted will probably need to take it very slowly, learning how to feel safe at each step. If one or both of you has been worrying about a sexual dysfunction, your comfort level will be much higher if you agree ahead of time to leave out areas that may trigger anxieties, most likely the genitals. Couples who have never done this kind of touching should begin with small steps, maybe just a back rub or a foot massage. Notice how it feels to be the toucher. Do you receive pleasure from giving? Or are you bored or busy evaluating how well you're doing?

After a break, switch roles and see how it feels to be on the receiving end. It's helpful if your partner asks what you like, but practice volunteering information by making pleasurable sounds when something feels good or requesting to be touched more firmly or gently, faster or slower, or in a different location. Relatively few people are easily proficient at this kind of direct communication. Even long-term couples find it difficult. Recognizing this fact can help you avoid yet another source of pressure: the need to be a perfect communicator.

Some couples complain that these experiments are too clinical. Anything new feels unnatural at first, but if you are willing to stay with it, soon self-consciousness will give way to playful fun. Exchanging pleasure-oriented massage is not only enjoyable, it's also a terrific way of learning about each other's preferences. No special training in massage is necessary. If, however, you wish to learn more about how to use the magic of touch to enhance your pleasure or to help solve problems, excellent books are available to guide you.[4]

CONFRONTING PLEASURE ANXIETY

It's common knowledge that too much anxiety inhibits pleasure. What isn't so well understood is that too much pleasure sometimes provokes anxiety. It should come as no surprise that children who are taught to mistrust their bodies grow up feeling uneasy with sensuous pleasures. Similarly, those whose bodies have been violated through severe corporal punishment or sexual abuse learn to think of their bodies as sources of pain rather than enjoyment. Others believe it's "selfish" to receive too much pleasure and therefore deflect touch away from themselves. Men and women who are self-conscious about their physical imperfections—and such feelings are difficult to avoid given the constant parade of perfect bodies in the mass media—may feel they're not sufficiently attractive to deserve extensive pleasuring or that no one could genuinely enjoy touching them.

Pleasure anxiety causes people to "numb out" because blunted sensations are easier to tolerate. One common strategy is to develop a rigid armor of tense muscles that act as an invisible shield against both pleasure and pain. Before anything can be done about pleasure anxiety it must be acknowledged with compassion and understanding. Once recognized, however, it's crucial to create opportunities to stop avoiding pleasure and learn how to relax and receive it. This is no small feat for those who are wary of the body's delights. But without a comfortable relationship with one's sensuality, the erotic adventure is ultimately doomed to be unfulfilling.

Nancy and Burt Revisited: Touching for self-discovery

When you met Nancy and Burt in Chapter 6 they had both stopped drinking but their sex life was in shambles. Nancy had realized that without alcohol she was too inhibited to act the sexual adventurer role she had perfected during the freewheeling 1970s. Instead a resurgence of guilt from her strict Catholic upbringing was turning her off completely. And Burt was wrestling with an old belief that he was boring and unattractive.

Having reached a point where they actively avoided all touch except friendly hugs and kisses for fear that they would repeat the cycle of expectation, rejection, and frustration, they were ideal candidates for sensate focus. They both liked the idea of pressure-free touching and eagerly made their first date for a nonsexual massage. Even though they both said they enjoyed the massage, neither showed much interest in trying it again. Week after week other responsibilities always took precedence. If anything, they touched *less* often than before.

People avoid touching experiments for all kinds of reasons, including a fear of raising expectations, the embarrassment of not getting turned on, the concern that one partner may feel more sexual than the other, or worries about not doing it right, to name just a few. None of these seemed to explain Nancy and Burt's avoidance. It was Nancy who first hinted at the truth. "I know this is supposed to be pleasurable," she said, "but I just can't seem to relax."

"Me neither," chimed in Burt. "You'd think I was a teenager on my first date. Back then I might have been nervous as hell, but at least I was determined to get inside her pants. I have no idea how it feels to touch for no reason other than it feels good."

When I asked them whether they felt more comfortable giving or receiving touch, both answered in unison, "Giving." Once their uneasiness about receiving became an open subject, they were willing to continue experimenting, especially when I encouraged them to pay close attention to what they were thinking and feeling. Like most people, they were under the impression that sensate focus is only successful for those who are able to ignore or suppress "mental chatter" and concentrate solely on physical sensations. Although it can be wonderful to do just that, watching random thoughts, feelings, and memories come and go is harmless, completely normal, and sometimes very instructive.

Nancy observed what professional body therapists know: the

most potent messages, memories, and emotions from childhood aren't simply stored in deep recesses of the brain, but are somehow etched directly into the tissues of the body. Once she began paying attention, she couldn't help noticing that the better Burt's touches felt, the more tense and guilty she became. "It doesn't matter what I think," she mused, "my body is convinced that pleasure is wrong, wrong, wrong!"—shocking words from a former sexual liberationist.

Nancy's insights emboldened Burt to observe himself the next time Nancy pleasured him. "When Nancy was stroking my inner thighs and balls last week," Burt explained, "I wanted to push her away even though it felt terrific. I guess I don't like being in the spotlight." By consciously resisting his urge to flee from the receptive position, he saw how he had learned to compensate for feeling unattractive by focusing on stimulating his lovers. Burt wasn't guilty about pleasure the way Nancy was. He just couldn't believe that anyone could genuinely want him to lie back and do nothing but enjoy. Burt also acknowledged that, like many men, he had trouble with the passivity of receiving pleasure. "Guys are supposed to be *doing* something," he said. "I feel like I'm slacking off on the job."

Nancy and Burt's inhibitions about pleasure were so thoroughly ingrained that only careful, courageous self-observation uncovered them. They took the risk of disclosing difficult, embarrassing feelings and attitudes they had concealed even from themselves. Gradually, the naughty thrills that once defined and limited their eroticism yielded to a pleasure-based approach. Sometimes they missed the heart-pounding intensity of overcoming so many obstacles on the road to passion. But the deepening bond between them brought a whole new dimension to their lovemaking.

As you can see, sensuous touch is a potent tool for healing, self-awareness, and simple enjoyment. It can also help you recognize hidden anxieties about pleasure, anxieties that aren't pleasant to confront but necessary if your eroticism is to evolve. Touch isn't the only way to come to your senses. Dancing, stretching, walking, or other forms of exercise offer opportunities to rediscover the joy of spontaneous movement and the vitality and strength that flow from it. Although these activities may seem far removed from sex, they all reconnect you with your body—the fountain of eros.

STEP 6:

RISK THE UNFAMILIAR

Therapists and clients alike often share a mistaken belief that gaining insight into the hidden roots of troublesome symptoms leads directly to more fulfilling behaviors. Although this notion holds an obvious appeal, it can also blind us to the complex realities of growth. Rarely is change automatic, no matter how insightful you are, especially when you're grappling with long-established patterns.

I'm not suggesting that self-awareness is useless, just that it promotes change only insofar as it emboldens you to try something out of the ordinary. Those who insist on spontaneous, "natural" change inevitably stick to the status quo—the only thing that truly comes naturally. Insights call your attention to the kinds of risks that are necessary. Then, if you can rise to the occasion, your courageous, *un*natural choices will yield far more results than any amount of armchair analysis.

RECOGNIZING AND SEIZING OPPORTUNITIES

There's a Buddhist saying: When the student is ready, the teacher will come. Sometimes that teacher is another person—a mentor, guide, or role model who challenges you to stretch your limits, while instilling confidence that you can to it. More often, life itself is the teacher, but only for those ready and able to follow an uncharted course. Whether we notice or not, life regularly invites us to step outside the constraints imposed by our habits. Unfortunately, most invitations go unanswered.

Whenever you make a mindful choice to respond differently than usual, combining awareness with concrete action, your goals and motivations make a leap toward actualization. Your decisions need not be dramatic to be effective. Even the smallest changes, when consciously chosen, make successive ones increasingly tangible and gratifying.

Carlos revisited: Allowing connection

Caught on the horns of a self-hating dilemma, Carlos, the unhappy voyeur you met in the last chapter, had learned in early adolescence to use inferiority as an aphrodisiac. Once he confronted how his CET magnetically drew him to idealized but unreachable men and dangerous situations, he faced many choices. For him, a major step was to disengage from locker room spying and venture into gay sex clubs, a world of consensual sex that sometimes leads to dates and occasionally even relationships.

Some people may have trouble seeing how engaging in casual sex could be a sign of growth, but for Carlos it was. He had a strict safe sex policy so he could explore his sexuality without life-threatening consequences. I remember his delight at spending hours with men who were as enthusiastic as he. Of course, the men whom he considered most perfectly masculine and frustratingly aloof invariably attracted him most. But there were other men who pursued *him,* a totally new experience he didn't know how to handle. Though it wasn't easy, he coaxed himself into not turning away. He knew that learning to accept positive attention was the only alternative to the searing pain of constant rejection.

Whenever Carlos was in emotional turmoil, however, his masturbation fantasies reverted to voyeurism. He noticed that his self-esteem took a nose-dive as he worshipped the objects of his desire. As a result of this awareness, he began to wonder what would happen if he were to celebrate his appreciation of the male form, to feel enriched by a man's beauty rather than demeaned by it. This became the central question of his therapy and his life.

Gradually, Carlos experimented—in both fantasy and behavior—with positioning himself more favorably in relation to men he admired. His initial discovery probably won't surprise you: inferiority was sexier. Equality, however, was infinitely more gratifying. The only people who can fully comprehend this distinction are those who know firsthand the intensity of eroticized self-hatred.

Carlos eventually began dating, cultivated the necessary social skills, and learned to avoid interpreting each man's reaction to him as a referendum on his worth. Eventually, he moved in with a guy who was handsome and affectionate but not as irresistible as the fantasy guys he continued to stalk in his imagination. This man wanted to love him, and Carlos's greatest achievement was the decision to receive that affection.

If you're pursuing any kind of erotic transition, large or small, I have a prediction. One day you'll look back on how far you've come and wonder how you did it. I doubt that you'll recall one pivotal turning point, except perhaps the decision to begin. Like Carlos, most likely you'll recall a series of unfamiliar and awkward experiments, opportunities seized in spite of—or because of—the risks. Through these experiments you turn the promise of change into a reality.

STEP 7:

INTEGRATE YOUR DISCOVERIES

If alternative erotic styles are to be more than passing experiments, they must become woven into the fabric of your everyday life. For a lucky few, this integration occurs with relative ease. Especially when modifications are readily satisfying and don't seriously conflict with older patterns, merely repeating pleasurable new behaviors may be sufficient to establish them.

For most people, however, it's not that easy. I've observed dozens of men and women attempt new forms of erotic expression, only to revert to less fulfilling but more predictable ways of being. They found that the most daunting aspect of an erotic transition wasn't trying new behaviors but making these changes stick. Unless you learn how to do this, your discoveries from the previous steps may, in the end, prove to be futile.

Few if any of us follow an uninterrupted path toward our goals. More commonly, we make headway until we come up against some internal or external blockage that tests our motivation. At such moments we may feel as if we're going backward. These setbacks are among the most common impediments to our development. But it's not the setbacks themselves that hinder growth. Far more important is how we respond to them. Some people quickly become demoralized, interpret their setbacks as signs of weakness, or perhaps even abandon their goals.

When you encounter a setback, I encourage you to view it not as a failure but as an opportunity to integrate change at a deeper level where it's more likely to take root and flourish. Sometimes this is simply a matter of recognizing reversals as thoroughly human and inevitable, and deciding to persist in spite of them—with as little self-criticism as possible. But I've learned from my clients that persistent setbacks serious enough to undermine positive change are, more often than not, signs that a person must wrestle with one or both of two crucial questions: (1) What is the relationship between the changes I seek and how I perceive myself? and (2) How do I deal with my old turn-ons as I practice new ones?

IDENTITY AND CHANGE

If you're determined to bring about meaningful changes in your erotic life, don't make the mistake of thinking of them in isolation. By its very nature eroticism interacts with your entire personality. Modifying it, even in seemingly small ways, often has unexpected implications. Some people aren't prepared for these ramifications and thus flee from change at the very moment when it's within easiest reach.

Identity—including your core beliefs—provides you with a sense of stability. Modifying how you act or perceive yourself may threaten that stability, so much so that you begin to question who you are. Most people find this ambiguity difficult to tolerate. Consequently, during periods of transition you may feel an urge to cling to the self-image you know best—even if this means slowing or abandoning your progress. The alternative is to take up the challenge of expanding or reshaping how you see yourself.

It's fascinating to observe the seemingly contradictory effects that growing has on identity. On the one hand, to grow is to become increasingly aware of and comfortable with your individuality. On the other hand, growing forces you to stretch in order to make room for previously rejected or newly discovered aspects of yourself. This enlargement of self is a hallmark of psychological development. Usually, however, it's also quite painful.

Regina revisited: Making room for love

Once Regina broke the stranglehold of silence about the sexual abuse inflicted on her by her stepfather, she was able to see that her need to seduce men had little to do with pleasure. She seduced to reassure herself that she was in control and to reaffirm the conviction that her value was as a sexual object. You may recall that when she instigated a crisis by slashing her wrists, she had recently begun dating a man who didn't fit the mold of the cold, aloof ones she usually chose.

No matter how much she expected him to use and abandon her, he refused. Instead he genuinely enjoyed her company, listened to her eagerly, held and caressed her passionately—and all without pushing for intercourse, which he sensed made her uncomfortable. One of the last mysteries Regina confronted in therapy was the irrational truth that this man's love had somehow pushed her toward self-destruction. The jagged cuts on her arms were symbolic reminders that she couldn't tolerate affection without suffering.

According to her long-standing core beliefs, Regina deserved exploitation and rejection. She was also trying to strengthen a more recent—but much weaker—belief that it was her birthright to be treated with respect. An inner battle was raging. Each time she accepted nurturance and love from Bob, she had to believe she was worthy of it. Some days that challenge was overwhelming and she would run away. "If he would just *use* me," she explained with extraordinary insight, "I would know who I am. It *hurts* to be loved!" Her pain was caused by the stretching of her identity.

To help you solidify positive changes, make a point of recognizing how each one coincides or conflicts with your self-image. Don't be surprised if you can't determine this easily. Live with new behaviors and ways of thinking about yourself for a while and notice how readily you become comfortable with them. If you find yourself avoiding the very things you think you want, consider this an indicator that your current identity is incompatible with the direction of your growth. It takes considerable courage to initiate the necessary shift in identity, but the process is quite straightforward if you decide to pursue it.

The most important thing is to continue exploring the changes you're having trouble integrating. But instead of avoiding discomfort when it arises, let yourself feel it, and eventually you'll learn about its source. Had Regina run away from her kind boyfriend as she often wanted to, she would never have confronted her internal reluctance to receive love. Only after she saw how her self-image was restraining her could she consciously begin to modify it.

How can identity be modified? Unfortunately, no easy answers exist. But I'm convinced that it's much simpler to *add* new self-perceptions to your identity than it is to banish old ones. In fact, if you struggle too fiercely with established beliefs, they will likely grow stronger. Rather than trying to get rid of or suppress self-defeating beliefs, you'll do much better if you concentrate on bolstering emerging beliefs that foster choice and self-affirmation. Sure you'll experience some inner conflict, maybe a lot. But as your new beliefs grow sufficiently strong, they will show you how to deal effectively with the old ones and reduce their damaging effects.

It also helps if you highlight aspects of your current identity that you can embrace wholeheartedly. Your identity is multidimensional and undoubtedly includes many features you can feel good about. Regina, for example, disliked her tendency to act as if she deserved mistreatment. At the same time, she appreciated her boldness, her willingness to take chances and, if necessary, to break the rules. She had always been a fighter, a characteristic that had helped her survive despite her antiself inclinations.

Recognizing, respecting, and expressing one's positive qualities stimulates an enlargement of identity and increases self-esteem. Because identity unfolds gradually as you grow up, altering it requires time and patience. Making room in your self-image for new potentials is a labor of love. Only those who decide they are worthy of that love can mobilize the necessary persistence. Then the expansion of identity becomes profoundly uplifting. One of the greatest joys comes from perceiving something old and familiar in a whole new light.

WHAT BECOMES OF TROUBLESOME TURN-ONS?

Almost as challenging as expanding one's identity is figuring out what to do with old sexual scripts as new ones take hold. In a logical world, problematic scenarios would quickly lose their appeal in the face of more fulfilling ones. In actual practice, most men and women who successfully create more gratifying turn-ons make a similar discovery about their erotic repertoires as they do about their identities: it's much easier to cultivate new sources of arousal than to cast aside old ones. This is the primary reason that I suggested in Step 1 to focus your goals on what you want rather than what you don't want.

Although it may go against common sense, integrating erotic changes normally doesn't involve obliterating problematic turn-ons altogether but rather finding a harmless place for them in an expanding, multidimensional self. Is such a feat realistic, particularly for those whose CETs have compelled them to reenact self-defeating scenarios? I'm convinced that it is. Needless to say, the ability to integrate once-destructive turn-ons into a self-affirming identity rarely develops easily. But unquestionably it does happen.

Ryan, the "prisoner of prohibition" who had spent most of his life struggling with his fascination with sleazy women, is a good example. With determination and courage he learned to enjoy warm, affectionate sex with his girlfriend, Janet. He even mourned the loss of the heart-pounding excitation his old activities once produced. Yet raunchy images continued to run through his fantasies no matter how much he tried to control them. For a time he couldn't resist declaring himself a failure.

It was obvious he had truly reached his goals only when he

learned to accept—and enjoy—these fantasies, without reactivating the inner battle that had made him miserable for so long, and without sinking into shame or self-reproach. Eventually, he even went so far as to use old naughty feelings, accompanied by a wisp of guilt, as harmless aphrodisiacs. Ryan's battle didn't end with the vanquishing of his old turn-ons as he once assumed it would. Instead his ultimate success stemmed from a transformation in how he used them.

Most people I've known who have successfully resolved troublesome turn-ons have been similar to Ryan. They gradually discovered that it wasn't their sexual scripts per se that had hurt them. The real problem was how they had learned to use their CETs against themselves. Once they stopped doing that—the most far-reaching change of all—they went on to develop more relaxed attitudes toward erotic material that was once deadly serious. By building an identity founded on self-respect, they used their imaginations to enjoy scenarios similar to those that once tormented them but without taking any of their detrimental aspects to heart.

PUTTING THE SEVEN STEPS TO WORK

Erotic transitions are as multifaceted as eroticism itself—full of detours and surprises. Try not to think of these steps as instructions to be followed like a recipe. To receive maximum benefit, approach them with a spirit of flexibility and creativity.

Concentrate first on the steps that capture your attention or evoke strong emotions. It is wise to consider any strong response, positive or negative, as a signal to look closely—if not now, then later. Don't worry about why you're drawn to one step or another. The key is to establish a high degree of personal involvement, and the best way to do that is to follow your natural inclinations. Later you can always take a second look at the steps that didn't particularly interest you the first time through. They may gradually take on new meaning as you grow.

These steps are synergistic—their combined effect is much more powerful than the effect of any single step. As you become engaged in any step, your involvement will have a ripple effect. Lessons learned from one step easily carry over to all the others. Also keep in mind that these are not the types of steps that require you to complete one before moving on to the next. Each step launches an odyssey that is never completely finished.

A WORD ABOUT AA AND THE TWELVE STEPS

If you're currently in a twelve-step program you may be concerned about whether the seven steps described in this chapter might conflict with your recovery. I've worked with dozens of people who have used the seven steps as beneficial supplements to the twelve steps of Alcoholics Anonymous or similar self-help groups. Many people in recovery have no idea how to recognize and cope with the erotic dimensions of sobriety, a topic rarely discussed at meetings. The seven steps are useful because they're specifically designed to promote sexual growth and healing.

If you're benefiting from a twelve-step program for a chemical addiction and also struggling with self-defeating sexual behaviors, you might wonder if programs such as Sex and Love Addicts Anonymous (SLAA), Sexual Compulsives Anonymous (SCA), or Sexaholics Anonymous (SA) might help. Although their social support and emphasis on spiritual renewal can be valuable, the disappointing fact is that these groups are among the least successful of all twelve-step programs. Many participants are discouraged by the unending litany of "slips" and by how few people have found comfort with their sexuality—so different from AA, where role models for long-term recovery are plentiful. Without becoming disillusioned about the wisdom of the twelve steps, it's important to understand why an approach that's so helpful for substance addiction may not be the solution for your erotic conflicts.

Erotic problems often involve compulsive repetitions and obviously have many features in common with chemical addictions, most notably an inability to modify behaviors despite negative consequences. Whereas recovering addicts completely sever the relationship with their drug of choice in order to find themselves, a person driven by sexual impulses can never sever the relationship with his or her eroticism.

The focus on abstinence that is appropriate for severe chemical addictions usually ends up making matters worse when applied to compulsive sex or obsessive "love." This is not to say that abstinence can't be a valuable tool as part of the process of erotic change. In Step 3 you saw that under certain conditions choosing to abstain from a problematic sexual pattern can be useful. But I'm convinced that a premature emphasis on abstinence increases the intensity of troublesome sexual urges by encouraging the struggle that fuels them. The erotic equation has shown us why fighting a sexual impulse only makes it stronger.

Alcoholics and other substance abusers usually have only one choice when it comes to their preferred drug—stay away from it. But no matter how compulsive sexual behavior may become, for erotic healing to take place, an increasing range of self-affirming choices must be claimed for oneself. One's power to choose is ultimately what induces sexual well-being.

WHAT ABOUT THERAPY?

All the people you've met in the last three chapters have worked with me in therapy. A supportive, nonjudgmental atmosphere facilitates disclosure of erotic secrets. In addition, some sort of therapeutic involvement is usually necessary to help uncover memories and beliefs that operate unconsciously. I'm not suggesting, however, that everyone with an erotic problem should enter therapy. Many people can make considerable progress on their own—if they are sufficiently motivated and know how to proceed.

Nevertheless, it's very difficult to probe the many layers of your erotic mind by yourself. It's a great relief to discuss thoughts and feelings honestly with at least one other person who genuinely listens without pushing any particular agenda. If you're fortunate enough to have a friend who listens respectfully and discloses intimate information of his or her own, perhaps the two of you can help each other with series of discussions.

Although many people consider their lovers or spouses their best friends, there's no set rule about how many of your deepest erotic yearnings you should reveal to your partner. No matter how intimate your relationship, your lover can never be truly neutral about all your turn-ons, especially ones involving other people. Many partners are also threatened by details about each other's past experiences. That said, I've known many couples who openly discussed the most private erotic matters and grew much closer—and quite stimulated—as a result. Especially when the time comes to try out new forms of sexual expression with a partner, your experiments are much more likely to be beneficial if the two of you communicate honestly about your intentions and feelings ahead of time.

Sometimes problematic erotic patterns are so much a part of who you are that you are unable to see them clearly—let alone resolve them. Then it may be wise to consult a therapist. But how do you know when to seek professional help? If two or more of the following statements apply to you, at least consider therapy:

- Even though I know something's wrong erotically I can't seem to define what the problem is.
- In spite of my best intentions I'm unable to initiate a change—even though I know what I'd like to accomplish.
- The attempts at change I do make don't seem to lead me anywhere.
- I'm uncovering disturbing memories or feelings that I don't know how to handle.
- I sense an inner conflict is sapping my energy but don't know how to call a truce.
- I continue to engage in certain sexual behaviors despite potentially damaging consequences.

It's not easy to find a therapist with whom you can work effectively. Believe it or not, many therapists, including some sex therapists, are uneasy about discussing the nitty-gritty details of eroticism. Interview several therapists and ask them to explain how they work with erotic problems such as yours. Be wary of therapists who seem to hold dogmatic beliefs about what healthy eroticism should be. Trust yourself and speak up if something doesn't feel right.

If you're already in therapy for other concerns, you may be reluctant to initiate discussions of erotic issues even if you suspect they are related to what you're working on. Therapists often don't inquire about your sexuality, so you might have to bring it up yourself. If certain parts of this book feel particularly relevant, discuss them with your therapist as a way of raising the subject.

Part III

POSITIVELY EROTIC

9

LONG-TERM EROTIC COUPLES

The creative use of learnable skills helps keep passion alive as intimacy deepens.

Nowhere are the paradoxes of the erotic mind played out more delicately, boisterously, and sometimes tragically, than in the crucible of committed, enduring relationships. One paradox, surely the cruelest, is that those couples who achieve the close, emotional connection that virtually all of us crave inevitably end up softening if not eliminating the obstacles necessary for passionate sex. Dr. Tripp describes this unwelcome reality:

> As the partners make the necessary compromises to achieve a high contact with each other, they win intimacy and a genuine closeness, benefits which contribute to the comforts of daily living and to their ability to get along with each other. For a time, their sexual compatibility soars as well. . . . But all this blending and complementation not only does not contribute to the lastingness of sexual attraction, it soon begins to dissolve it. Thus, in well-balanced ongoing relationships, the compatibility of the partners tends to progressively improve while their erotic zest for each other markedly declines.
>
> Conversely, a genuine closeness is notable by its absence in the most intense forms of erotic interest, including the high romance that can so disconsolingly occur between

> utterly mismatched partners. Thus it is in new relationships and in marriages torn by fights and clashes—that is, where complementations and details of compatibility have *not* been worked out—that the highest erotic excitements flourish.[1]

According to the ideals of love and marriage to which most of us subscribe, deepening affection and closeness are supposed to coexist with a dependable, satisfying sex life. However, the difficulty millions of couples have in combining closeness with sexual enthusiasm is evident in the steady stream of books and articles about keeping the spark in marriage. Marital experts often insist that waning passion is the result of poor communication skills, a lack of intimacy and trust, or unresolved conflicts. While it's true that any of these can lead to unsatisfying sex, it is most definitely not true that good relationships automatically lead to good sex. In fact, my observations match those of Dr. Tripp. It is often in the best relationships that passion becomes most elusive.

I'm also convinced that couples who openly confront the difficulties of combining intimacy and passion are the ones most likely to thrive. It is crucial to acknowledge that closeness and sexual desire are not one and the same, but rather two separate, yet interacting experiences. Their rhythms vary tremendously according to how each relationship begins and unfolds.

THE EVOLUTION OF SEXUAL RELATIONSHIPS

Although many have attempted to codify the stages of committed love, the fact is that no two partnerships follow exactly the same course.[2] Nor is it possible to predict with confidence which couplings will endure and which won't. I'm sure you've known partners who appeared to detest each other yet stayed together anyway. Conversely, other couples who seem genuinely to care for each other surprise everyone when they separate.

Love refuses to conform to rational notions about how it should begin, progress, flourish, or die. Nonetheless, the vast majority of couples I've worked with grapple with similar turning points in the development of their relationships, each with implications for the quality of their interaction—both in and out of bed.

THREE PATHS TO COUPLEHOOD

The emergence of a couple marks the birth of an entirely new entity. In the formation of *we,* both the self and the other are changed. Sometimes two people know from the first shared glances that their lives are destined to be intertwined. For others the sense of *we-ness* that eventually forms the nucleus of a lasting bond builds slowly with innumerable ups and downs. Not only do couples form at different speeds, they also adopt different interpersonal styles based on what draws them together:

Passionate couples are swept up in the intensity created by the dance of opposites.

Companionate couples are founded on mutual understanding, resonance, and comfort.

Pragmatic couples are concerned with practicalities such as availability, money, prestige, or social acceptance.

To some degree, each couple is a blend of styles, but always with a distinct emphasis.

Because passionate connections spring from the magnetism of contrasts, they are the most likely to be cemented, at least initially, by rushes of lusty heat combined with the thrill of limerence. The lovers feel so vibrantly alive and whole that practical questions of compatibility become irrelevant. Consequently, passionate couplings can be fragile, often unraveling as dramatically as they began. But passion can also launch two individuals onto an evolving lifelong odyssey. I'm sure that far fewer of us would ever take the plunge into couplehood without the unbridled passion that helps overcome fear and reticence long enough for a genuine bond to take hold.

The passion produced by contrast is one kind of chemistry. Another kind—equally powerful, but less frenetic—occurs when companionate couples bask in the joy of profound similarities and mutual resonance. These partners typically describe each other as soul mates or best friends. Whereas passionate lovers may use the terminology of friendship to express feelings of oneness, they sense their relationship is quite different from other friendships, a fact their friends definitely notice.

Companionate lovers really *do* feel and act like friends. Most of them enjoy sex, especially at first, but recognize the absence of the high intensity they may recall from stormier past relationships. For a time this doesn't matter because their commonality brings a beautifully quiet passion to their lovemaking. They're often amazed by how easy it is to be together and quickly settle into a high comfort level with minimal insecurity.

In most cases, an awareness of similarity is friendship, whereas an awareness of difference is passion, so I'm sure it won't surprise you that while companionate relationships are often highly romantic, some aren't particularly sexual. A significant portion evolve into friendships, with one or both looking for passion elsewhere. Those who stay together, however, often become highly compatible mates.

Some practical aspects of pair bonding aren't at all romantic. Pragmatic couples are primarily brought together by convenience or availability, even though neither partner stirs deep passions or elicits a special resonance in the other. Other practical considerations include the ability to be a good provider, reliable mate, or trustworthy parent. Because pragmatic connections begin in the mind rather than the heart—often accepted or rejected after a careful weighing of pros and cons—they tend to be short on magic and zest. In their most extreme form, these relationships have the flavor of a business deal.

Despite their limitations, some pragmatic pairings have amazing staying power, in part because they're relatively free of the grand expectations and fantasies that often set other couples up for disappointment. And don't forget that, like all relationships, pragmatic connections evolve. Although it's quite rare for passions to flare where none existed at the beginning, this does occasionally happen. But it is far more likely for pragmatic couples to develop genuine bonds of affection and respect.

RUDE AWAKENINGS

Once two individuals think of themselves as a couple, they begin to find out who, exactly, this fascinating other really is. In a state of high limerence we usually think we know a lot more about our partners than we actually do. Especially if a connection springs from pure passion, much of what we think we see may turn out to be a projection of our fantasy image onto the beloved. Yet passionate attractions are rarely *all* illusion. We also see in the beloved potentials yet to be realized. Love's eyes look beyond everyday reality, which can be a blessing as well as a curse.

Because the joy of new love brings out the best in both lovers, many of their early discoveries are happy confirmations of their original attractions. Eventually, though, there will be rude awakenings. Occasionally, distressing secrets come to light. More often there is a gradual realization that characteristics which originally appeared positive also have a negative aspect. For a time Ben loves Melody's tendency to be absentminded because it reminds him that she is preoccupied with deep thoughts—a quality he admires. Whenever she forgets something, she apologizes with puppy-dog eyes that make Ben melt. A couple of years later when they own a house and Melody consistently misplaces important papers, his patience wears thin. Similarly, Larry is at first strongly attracted to Mark's confident masculinity and independence, only later to realize that Mark is completely out of touch with his feelings and walks out of the room whenever Larry tries to discuss a problem.

Rude awakenings, and the conflicts that inevitably follow, provide opportunities for the partners to deepen their attachment. Problems arise when lovers become locked in life-or-death power struggles. Partners who once perceived each other as perfect, precious gifts begin to demand changes. When appreciation is overshadowed by criticism and complaint, few people admit how perplexing and hurtful it is to receive disapproval from the very person who once made them feel miraculously whole and lovable.

Many of the rudest awakenings aren't as much about the other person as they are about yourself. The same love that makes you feel terrific also activates deep vulnerabilities and insecurities. The beloved turns out to posses an uncanny ability to push your buttons, to stir up difficult emotions from the past. You find yourself acting childish or parental even when you don't want to.

You begin to notice an annoying sensitivity in your partner, who reacts way out of proportion to your words and actions. Good intentions seem to count for nothing. In the most severe cases partners become so highly reactive that almost any discussion leads to fights and misunderstandings. Both feel under siege, which in turn activates primitive defense mechanisms that make matters infinitely worse.

Almost any rude awakening can undermine a couple's sex life, particularly when conflicts and resentments accumulate and fester. Of course, some of the rudest awakenings involve sex directly. Almost every couple encounters differences in desired activities, frequency, or timing. Although there may have been subtle hints from the beginning, even serious incompatibilities can be easily concealed by the intensity of new love, especially in relationships founded on fiery passion. Lovers on a romantic high have an uncanny ability to satisfy each other even when they later turn out to be seriously mismatched. As time passes and passions cool, disparities that once seemed inconsequential loom large.

TWO FACES OF COMMITMENT

At some point most couples openly announce or quietly assume that they've become committed. Couples are usually not in as much agreement about the details of their commitment as they assume, but common expectations include spending regular time together, honoring plans and promises, being available during times of need or trouble, and maintaining sexual exclusivity. Commitments provide security and thus can facilitate sexual pleasure by calming fears of abandonment and rejection. But for those who enjoy the hunt, the thrill of the new, or the focused concentration of courting and being courted, security can quickly become antierotic. Some may associate commitment with being trapped. Rather than craving to be with the beloved, they start to dream of escape.

Commitments inevitably involve obligations, but when commitment and obligation become indistinguishable, the stage is set for serious sexual trouble. Compelling erotic desires are always "want to's," whereas obligations are "have to's." When freely made, commitments are powerful statements about wanting to sustain a connection and are therefore fully compatible with desire. By contrast, obligations rarely call forth anything more than a grudging willingness to meet them. Partners who perceive their commitment primarily as a set of obligations are heading down a slippery slope that ultimately leads to the dampening of desire.

SETTLING IN

Couples who weather their rude awakenings and voluntarily renew their commitments despite differences and disagreements typically enter a period of relative calm. They disengage from pointless power struggles as each increasingly takes responsibility for his or her own irrational reactions. Gradually, they learn to see conflicts as opportunities for self-discovery and creative problem-solving.

As they move from conflict to cooperation they redirect some of their energies into projects such as career development, home acquisition and nesting, or conceiving and raising children. During this period many couples also solidify a network of mutual friends and establish rituals such as annual events, vacations, and celebrations. For most couples the early years of settling in are among the most satisfying.

Eventually, however, as the frequency and intensity of sex begin to taper off, signs of sexual cooling become impossible to ignore. In most cases this happens gradually over a period of years. However, if one partner, most likely a man, has trouble feeling love and lust toward the same person, his desire may plummet very quickly after the intensity of early limerence calms. For these men—and some women too—genuine closeness is a complete turn-off.

But even among couples who are capable of enjoying the interplay of love and lust, two opposite forces contribute to a progressive reduction in sexual enthusiasm: (1) boredom and emotional disengagement, and (2) increasing closeness, familiarity, and comfort.

Most couples develop sexual routines. For some these routines are acceptable—even comforting—especially if they are punctuated with occasional surprises. Others are so bored by routine that they find it increasingly difficult to generate sufficient sexual energy to become highly aroused or have orgasms. Without noticing, many also drift apart emotionally. Not only do they spend less time together, they talk about practical matters—such as what to do about the kids, problems with the house, or finances—with fewer of the heart-to-heart discussions that once brought them closer.

As they settle in, other couples become increasingly intertwined. Some grow so close and comfortable that they act more like siblings than lovers; sex might even feel a bit incestuous. Also, the same intimacy that makes for a wonderful connection can obliterate the last vestiges of desire-enhancing obstacles. They become so close that the chemistry between them is neutralized.

Cooled passions, whether a result of emotional disengagement and boredom, or closeness and comfort, sooner or later become a challenge for almost every couple. If their original bond was based on passion alone, they may miss the intensity so terribly that they terminate the relationship. Obviously, passionate couples are more likely to survive sexual cooling if more than sex holds them together.

Companionate couples are typically among the first to feel the loss of genital arousal, partly because they felt it least to begin with and partly because they establish comfort so quickly. It is difficult to predict how these couples will respond. Some who never shared high erotic intensity, and continue to enjoy the simple pleasures of sensuality and affection, aren't particularly troubled. Others face a serious crisis, especially if one or both develops a sexual dysfunction or loses interest in sex altogether.

Although hardly anyone welcomes it, many couples respond to sexual cooling with humor and grace. In my experience, these are the ones who recognize the reduction of erotic zest as a natural occurrence that calls for creative adaptations and adjustments. The secrets of their success will be our focus in the next section.

THE REEMERGENCE OF THE SELF

In marital lore "the two become one." In actuality *becoming* one can be very sexually stimulating, but *being* one isn't—at least not for long. In every long-term relationship one or both partners eventually feels the tension between me and we. There may come a time when the balance shifts too far toward mutual interests to the exclusion of individual ones. At such times it can become crucial for the survival of the relationship, as well as the nurturance of sexual desire, for them to focus on individual needs.

Part of this process involves disentangling, a clarification of the psychic boundaries that often become fuzzy during couplemaking. Entanglements occur when the feelings and needs of one become confused with those of the other, or when one chronically shapes his or her words and actions according to expectations about how the other will respond. These potentially suffocating entanglements are often mistaken for intimacy—a dangerous misconception. Togetherness and merging are very effective for cementing a bond, but true intimacy takes two separate individuals who are willing and able to balance the dual imperatives of individuality and interdependence.

Reasserting the right to be separate may trigger an upsurge in disagreements and fights. These will be particularly intense if one partner feels abandoned by the other, or fears that the partnership is losing its reason for being. If one partner's desire for more separateness draws out a clinging dependency in the other, they may slip into rigid roles of "distancer" and "clinger," unwittingly pushing each other into more extreme positions.

Despite difficulties, the reemergence of the self often works wonders for a couple's sex life. When partners expand their range of interests and become more engaged with and stimulated by the outside world, their attraction for each other typically grows. Not only are they less able to take each other for granted, some also discover that insecurity about what the other might be doing actually heightens their interest. Ultimately, the reemergence of the self furthers the long-term goals of the couple, as long as the needs of the individual can find a way to coexist with an ongoing devotion to the partnership.

THE SEASONING OF EROS

Financial crises, natural disasters, major illnesses, family conflicts, deaths of loved ones, affairs—these are just a few of the difficulties that send shock waves to the very core of even the best relationships. Sad are the pairings that collapse under the strain. Fortunately, many couples discover that when their love is tested and survives, it grows and deepens. Love that flourishes under duress establishes its right to exist. Seasoned lovers learn to tolerate the other's quirks and foibles without seeing them as personal affronts. Happier couples go further still: they grow to accept the other's eccentricities, even the annoying ones, as part of what makes the beloved unique.

Seasoned lovers, sometimes after years of struggle and turmoil, eventually achieve minimal conflict between their shared bond and their individuality. Their intimacy facilitates rather than interferes with each individual's growth. Theirs is the "being-love" Abraham Maslow wrote about, as opposed to the far more common "deficiency-love."[3] Whereas deficiency-love is motivated by the need to fill in or compensate for one's missing aspects, being-love is a union of self-acceptance and the capacity to love. Those who find their way to being-love are blessed by simultaneously feeling at home with oneself and the beloved.

And what becomes of passion as love matures? It changes, to be sure. Sometimes it fizzles—but by no means inevitably. Many couples never lose sight of a remnant of their original passion, although they experience it in less boisterous forms. They recognize that eros, if it is to survive the ravages of time, familiarity, and routine, requires a special kind of nurturing and a unique set of skills.

SKILLS OF EROTIC COUPLES

It's easy to see how enduring coupledom can undermine both attractions and obstacles, which—according to the erotic equation—are two key ingredients for passion. Yet there's plenty of room for optimism. All across the land, creative couples are maintaining fulfilling sex lives for ten, twenty, thirty years or more. In the early 1980s Blumstein and Schwartz studied thousands of couples in the United States, including married ones, straight cohabitors (not married), and lesbian and gay couples.[4] Although the frequency of sex dropped over time for all types of couples, the vast majority continued to enjoy it regularly. Interestingly, married couples tended to remain the most active, with almost two-thirds reporting some sort of sex at least once a week, even after ten or more years together. Long-term gay and lesbian couples had sex together less often, but most still did so at least once a month.[5]

What fascinates me is how long-term couples nourish a vital spark of eroticism, despite all the factors that can conspire to reduce it. Sadly, we know far too little about this. For one thing, couples rarely discuss the details of their sex lives—often not even with each other. When they are willing to talk openly, as they sometimes do with sex researchers or couple's therapists, their beliefs about how they sustain sexual interest may be quite different from how they actually do it.

Erotic couples regularly employ simple tricks to invigorate desire, techniques that could be provocative or hurtful to their partners if they were revealed. For instance, more than a few committed men and women engage in harmless flirtations with others. If their partners were to witness the flirting, some would become jealous. But many of these partners reap the benefits of having highly stimulated lovers, without having to confront all the reasons why.

Then, of course, there are the couples—nobody knows how many—who go through the motions out of a sense of duty, though they receive minimal pleasure from it. They view sex as necessary maintenance, like doing laundry or grocery shopping. Some of us might consider this an unacceptable solution to sexual boredom, but for many couples it works.

Most of what I know about long-term couples I've learned while doing therapy with individuals who discussed their relationships or in joint sessions with both partners. Of course, it's useful to hear about the sexual concerns that motivate people to seek help, but it's especially valuable to see how couples face their problems and take effective actions to make things better.

I've also been enlightened by working with couples who entered therapy because of nonsexual problems. Lots of couples maintain active and apparently satisfying sex lives in spite of chronic disagreements and fights. In a logical world, the quality of a couple's erotic life would bear some relationship to the quality of their overall relationship. But I've worked with a number of couples who learned to cooperate more and fight less only to discover that their sex lives got worse. They apparently needed upheaval to fuel their passions.

Theories about how couples *should* behave sexually are of little value and often do great harm by setting up unrealistic expectations and distracting the couple from the delicate adjustments, compromises, and inspirations that have the best chance of working. What interests me is what long-term couples actually do, in concert or individually, to keep sex satisfying as they develop other positive aspects of the relationship—such as companionship, mutual support, and, in the most highly evolved relationships, a loving commitment to the other's well-being and growth.

In this section I want to call your attention to a set of crucial skills that erotic couples appear to develop and apply with remarkable regularity. Long-term couples who discuss their sex lives with me always end up focusing on several—sometimes all—of these skills in one form or another. How consistently couples recognize them and put them to good use appears to have a major impact on the quality of their sexual interaction.

I'm quite certain that the majority of couples who conscientiously cultivate these skills will benefit. Nevertheless, when sexual interest is waning there are no sure-fire solutions. The great issues of erotic life are so much a part of the human adventure that they can't be fixed like a leaky faucet; they can only be lived. In the living, possibilities emerge that defy logic, sometimes with happy results, sometimes not. Whether you are preparing for a future relationship or already involved in one, your journey will be infinitely more satisfying if you realize that the unfolding of eros is a work in progress that is never finished.

REFINING THE ART OF EROTIC COMMUNICATION

Sex therapists and educators are forever harping on the value of communication. To sustain satisfying sex lives, partners must exchange a tremendous amount of information, much of which changes with time and circumstance. Most couples rely primarily on nonverbal clues about how the other likes to be approached and touched. Yet even the best nonverbal communication is severely hampered because it depends on trial and error, with feedback limited to ambiguous groans and squirms.

Hardly a week goes by that I don't encounter a couple who discovers in therapy that long-held beliefs about the other's likes and dislikes are completely inaccurate. Such mistaken convictions typically are solidified when subtle signs are either misinterpreted or never tested again. During one session Marty described with matter-of-fact certainty how Debbie, his wife, didn't like him to pump vigorously during intercourse. "Where did you get *that* idea?" Debbie asked with a look of utter disbelief. "A couple of times you've pushed me away and you sure *looked* unhappy," he retorted. "I was," said Debbie with exasperation, "but that's because I wasn't wet enough. When I'm hot and lubricating I love it when you get rough!" Marty and Debbie had a genuine desire to please each other. But you can see how *only* words could describe the specific conditions under which Debbie welcomed vigorous intercourse.

No matter how sophisticated we think we are, the fact is that talking about sex is difficult for everyone, even those of us who comfortably bring up sexual matters professionally. Not only are sexual desires, worries, or frustrations hard to articulate, but your partner holds special powers to hurt you by poking fun at you, ignoring you, or acting as if you're weird. No wonder productive discussions about sex are so difficult to initiate.

Sexual communication has a dual function. Besides passing along crucial facts about what you like, don't like, or might like to try, it's also important to bolster—or at least not undermine—your partner's sexual self-esteem. There's no question that we're at our best when our partners make us feel as if we have a special knack for turning them on. Conversely, we're our *un*sexiest when we're convinced we can't do anything right.

Don'ts and dos of sexual communication

Most couples know when communication is necessary but aren't sure how to proceed. To that end I offer some suggestions. Because some errors are virtually guaranteed to turn a discussion into a disaster, I'm going to start by violating a truism of marriage counselors that absolute words such as "always" or "never" be judiciously avoided. No matter how tempted you may be, never do these three things:

First, never belittle, berate, or hurl sexual insults at your partner, no matter how upset you are. Damaging your sex life is too high a price to pay for the momentary satisfaction of lashing out. It's much better to talk sincerely about how you feel, even if you must wait until you calm down. Sexual insults are very difficult to take back. So even if you're an ardent believer in speaking your mind, stifle yourself whenever you feel an urge to attack your lover's sexuality.

Second, never compare your partner to past lovers. Not only is this downright rude, it makes it harder for your partner to let go of inhibitions and un-self-consciously reveal his or her sexuality to you—a prerequisite for maximum satisfaction.

Finally, never complain about your sex life to a mutual friend. Not only does this place your friend in an awkward position, but such complaints have a way of coming back to the partner, often exaggerated and distorted. It's perfectly valid to talk with a friend about your problems, but pick someone who doesn't have a close connection with your partner. In addition, request an explicit promise that your disclosures will remain confidential.

Those are the don'ts. Fortunately, there are also do's that help make talking about sex easier and far more productive.

First, make a habit of regularly giving your partner positive verbal and nonverbal feedback about what feels especially good. This builds his or her confidence and increases the chances that you'll receive more of what you like. What psychologists call positive reinforcement (praising what feels good) is much more effective at bringing about change than punishment (complaining about what's wrong). Let's say you wish your partner would stroke your inner thighs with gentle caresses rather than grabbing you like a hunk of meat. It's perfectly valid to say, "I like it best when you touch my thighs gently like this," and then either describe your desires or, better yet, take your partner's hand and demonstrate. You'll get even better results if you also whisper, "Ohhh, *that* feels wonderful" the next time your partner touches you even a little bit more like the way you enjoy it.

Second, discuss sexual problems and dissatisfactions when you are feeling close, not when either of you is defensive, tired, or preoccupied. Some of the best opportunities are during or after any shared pleasurable experience. The goal is to create a link in both of your minds between feeling good and talking about sex.

Third, if it's too difficult to talk directly, write your lover a note. This technique allows you to refine your message and gives your partner plenty of time to digest it without reacting. Rather than assuming your motives are obvious, describe them explicitly—for example, "I value our sexual relationship so much that I've been thinking about how to make it even better."

Fourth, learn to be a good listener, which requires that you receive information without becoming defensive—no small feat when you're discussing material that touches on your most vulnerable areas. You probably know what it's like to appear to be listening even when you're not. Especially if you're feeling criticized you may be so busy preparing your rebuttal that you miss the essence of what your partner is trying to say. Sometimes it's better to separate speaking from listening, two distinct yet interacting skills. As an experiment, let one person speak for five minutes while the other just listens. Then let the speaker become the listener.

It helps to inquire specifically about what your partner is experiencing and how you might facilitate his or her pleasure. The goal of these delicate questions is to give your partner permission to express himself or herself, without sounding as if you're demanding something. You might try touching your partner in a couple of different ways and places and asking which feels best. Vague questions such as, "What turns you on?" are likely to bring you evasive rather than enlightening responses.

Fifth, become fluent in the language of emotions. Emotional expressions are not only associated with vitality and passion, they also counteract boredom and increase intimacy. The trouble is, one partner usually wants them more than the other. Often, but not always, men are uncomfortable with emotions—except possibly anger—because a lifetime of masculine socialization has taught them to perceive emotions as signs of weakness. But remember that personality differences also play a part. After all, even in same-sex relationships one partner is typically more feeling-oriented than the other.

The irony is that the person who wants more openness with feelings tends to increase his or her own emotionality. Unfortunately, this only widens the gap between the partners when the feeling-oriented one inadvertently confirms the reluctant one's fears that feelings will get out of control. As a result, he or she becomes even less emotional.

Sixth, know when to keep quiet. Too much discussion is the enemy of communication, though couples often disagree about how much communication is too much. Ironically, when one partner wants to talk more and the other less, they often end up driving each other to even more extreme positions. "Why do you want to talk constantly about our relationship?" one partner demands. "Because you never do!" retorts the other.

Couples with a significant disparity in their desire or tolerance for intimate talk can sometimes break this impasse by deliberately switching roles. The partner who wants more discussion agrees, as an experiment, only to bring up difficult topics at predetermined times, while the one who usually clams up agrees to try initiating. But note: this approach only works if both partners freely accept it. Talkers who unilaterally test their partners by shutting up to see if and when they will say something are playing a dangerous game with predictably negative results.

Finally, one of the most effective ways for couples to expand their knowledge about what turns them on is to remind each other of past encounters together that were particularly pleasing. There's only one pitfall to avoid. If your reminiscences come primarily from an earlier, more passionate phase in your relationship, you might set yourselves up for unfavorable comparisons with your current sex life. Make a point of emphasizing specific recent encounters. With rare exceptions, even couples who think their sex lives are a complete mess can point to at least *some* pleasurable moments, thereby highlighting the fact that they have a positive foundation on which to build.

CULTIVATING WARM SEX

My observations have consistently revealed an apparent contradiction: to preserve opportunities for lusty, passionate sex, most successful long-term couples develop the ability to enjoy "warm sex." Rather than emphasizing focused intensity, warm sex revolves around calmer experiences of sensuality, affection, pleasure, and playful fun. Although warm sex usually includes genital stimulation, its goals are neither high arousal nor orgasm.

Especially during periods when desire is relatively low, warm sex allows couples to maintain a physical bond and helps them to continue perceiving each other in a sexual light. In my work, I have regularly been reminded of how crucial this is. In fact, I haven't seen a couple—nor have any of the colleagues I've informally surveyed—who were able to rebuild a sexual connection after they had stopped thinking of each other in an erotic way for five or more years.

Some people have great difficulty learning to enjoy warm sex, while others gravitate to it naturally. As a sex therapist I often encounter men and women who think of warm sex as boring sex—or no sex at all—and therefore avoid it or become frustrated when it doesn't always lead to intercourse. Men are often troubled by warm sex because it doesn't necessarily generate sufficient erotic energy to produce or sustain an erection. Unfortunately, heterosexual and gay male couples typically think the firmness of the penis is *the* indicator of how well things are going, an idea inimical to enduring sexual satisfaction.

In addition to its other rewards, warm sex maintains an erotic playground. The passion of new love is typically an automatic catalyst for sex. As the urgency calms, however, seasoned lovers make time for sex, an obvious fact proclaimed so often by marriage counselors that it now borders on cliché. But if you want to sustain a happy sex life over time, this is one cliché you can't afford to ignore. When erotic couples evolve out of the heat of limerence, it's crucial that they find ways of keeping sex a priority—not just in concept but in fact.

Some parents feel guilty if they take time away from the kids to focus on their relationship. Couples need to remind themselves that their connection is the core of the family and cultivating it is, in the long run, one of the best things they can do for their children. By the time kids are able to entertain themselves with a minimum of supervision they should also be able to grasp the concept of Mom and Dad having private times when they are not to be interrupted (except, of course, for emergencies). A lock on the bedroom door is essential. A playful sign such as, "Do not disturb. Mommy and Daddy snuggling" is not only a practical help but also communicates positive messages about sensuality and affection—the best sex education there is.

SUSTAINING AND BUILDING ATTRACTION

One of the key challenges of long-term loving is how to keep attraction alive. Attraction tends to be stimulated most strongly by the new, the unfamiliar, or the unattainable—all features that decrease with day-to-day living. How then do erotic couples maintain their mutual appeal? Years of observing and questioning have led me to conclude that, while no simple answers exist, there are two basic strategies. The first is to stay in touch with the original attractions that brought you together in the first place. The second involves recognizing new sources of attraction as the relationship evolves. Most erotic couples rely on a combination of both.

Regardless of the specific problems that bring them to therapy I always ask couples what first drew them to each other. Not only does the ensuing discussion provide me with valuable background information, it also gives them an opportunity to focus on crucial memories that may have been overshadowed by more recent conflicts and concerns.

Erotic couples make a point of remembering their original attractions because they realize that even small remnants of these attractions can be powerful aphrodisiacs. For instance, passionate lovers focus on specific physical features that continue to be arousing, even when both of their bodies may have seen better days. This is a positive manifestation of the lusty objectification many people find objectionable.

Erotic couples are also aware that attraction is inspired by much more than the physical, so they also pay careful attention to the behaviors and attitudes that turn them on. At least as important as recognition of attractive features is the ability to bolster your partner's sense of attractiveness by making affirmative statements about him or her and to accept and value your partner's perceptions of you.

Thelma and Max: That look

I once had an opportunity to discuss the interplay of mutual attraction with a delightful couple who had been married for almost forty years. Thelma, a feisty woman in her mid-sixties, and her husband, Max, who had just recently turned seventy, consulted me for advice on how to handle a bitter custody dispute their son was involved in following his divorce. The entire family was in an uproar. However, Thelma and Max's relationship was doing just fine. When I asked about their secrets of long-term love they were embarrassed at first but soon began talking about their enduring attraction.

"Sometimes I think I'm all washed up," Max explained as he reached for Thelma's hand. "But this little lady just gives me that look and I feel twenty years younger."

"It's not me," Thelma protested. "I don't do anything. I can't help it, he still looks just like the man who swept me off my feet—so suave and handsome, and just as confident as can be."

As I listened to Thelma and Max I couldn't help noticing that their abiding attraction was totally interactive. With her eyes Thelma communicated her admiration to Max, who responded by feeling more attractive. As a result, he exhibited more of the qualities that attracted Thelma, who in turn felt validated by having the attention of such a desirable man.

Thelma and Max also reminded me that the ability to sustain

attraction is, to a significant degree, a conscious decision, an act of will. The idea that we can choose to feel attracted goes against the notion that attraction is something that happens to us. Fresh attractions often exhibit this effortless, automatic quality, as though you're being made to take notice. Ongoing attractions operate differently. Sure you want to be moved by your partner's desirability. But it is your responsibility to be actively receptive, to open your eyes enthusiastically to the beauty of your lover. Attraction's biggest enemy is the tendency to stop paying attention. There is no doubt that invisible partners are never attractive.

One of the most difficult problems is also one of the most common: what to do when one or both partners have "let themselves go." Few are comfortable saying—and even fewer hearing—such comments as, "You're getting fat and unattractive." Yet people often tell me in private how distressed they are when their partners act as if they're entitled to attraction no matter how little they do to maintain themselves. As difficult as it is to discuss such matters, I believe it is essential. To remain silent as you watch your lover give up on his or her attractiveness is to tacitly agree that how he or she looks no longer matters.

It's not always possible to retain remnants of an original attraction. The characteristics that first bring partners together can change radically or disappear completely over time. Some couples respond by turning their attention elsewhere. Erotic couples, however, make a conscious effort to develop new sources of attraction as old ones fall away; they learn to see their lovers through new eyes.

Richard and Mark: Changing faces of attraction

Having lived together for over eleven years, Richard and Mark had been through their share of trouble, including one extended separation, yet their commitment remained solid. Mark consulted me because he was concerned that his attraction for Richard was fading. Like most male couples, after their first few years together they had agreed to allow casual sex outside the relationship. There had been mutual insecurities during the transition to nonmonogamy and two crises when each had a brief affair, but their arrangement had mostly worked well. In addition to their outside activities, they also had continued an active sex life together until the last year.

"I don't get it," said Mark, "Richard is still very handsome. He has a beautiful body. We have lots of fun together. But I feel myself turning off and it scares me."

As usual, I asked what had originally attracted Mark to Richard. He immediately mentioned Richard's youthful good looks, playful sense of fun, and flair for adventure—all of which contrasted with Mark's more serious, conservative nature. "Even as a child," Mark explained, "I was a little adult, dutifully doing chores, schoolwork, and taking care of my sister. Richard knows how to have fun and even gets away with being irresponsible—or at least he used to."

Over the years Richard had changed dramatically. He achieved an advanced degree, established a new career, earned more money, became unmistakably more independent and powerful. Mark felt proud of Richard's accomplishments, but the contrast that had excited him for so long was narrowing to practically nothing. Now they were both dominated by their responsibilities. As a result, Mark had been noticing younger, less mature men—more like Richard used to be.

Mark hadn't given these changes much conscious thought until his discussion with me. After that he talked with Richard, who soon joined us in the sessions. He too had lost some of his original attraction but in a different way. Whereas Mark still was drawn to the youthful ways that balanced his seriousness, Richard was no longer interested in big brother types. He wanted Mark to be attracted to him as a fellow professional and peer. "I want to be two men together," he proclaimed, "not a man and a boy."

When couples speak of "growing apart," they are often referring, without realizing it, to changes in one or both partners that have undermined their original attractions. Like Mark, many people find it difficult to adjust their attractions to match new realities. Mark and Richard found their open but gentle discussions of these matters helpful. As they became increasingly honest with each other, they felt closer and Mark's sexual interest increased somewhat.

Before leaving therapy Mark tried a bold experiment. He decided to try being more playful and adventuresome himself, instead of looking for these qualities in younger men. Though it wasn't easy, he found that even small displays of youthfulness on his part helped him appreciate Richard's new level of maturity and confidence. It also made him realize that his lifelong lack of opportunities to feel young (even as a child) would probably always pull him toward men whose boyishness could help balance his one-sidedness.

Just because strong patterns of attraction tend to be stable doesn't mean they must remain static and inflexible. Erotic couples accept both the stability of old attractions and the risks of new ones. What matters most is that, like Richard and Mark, even as they flounder they remain committed to their partnerships.

COPING WITH MONOGAMY AND ITS ALTERNATIVES

Most people who are strongly attached to their partners feel insecure simply imagining the beloved having sex with another. Ironically, the declaration or assumption of monogamy often gives forbidden objects a special allure. Outside erotic interests take three different forms: flirtations, dalliances, and affairs.

Flirtations are the mildest sexual interactions outside the primary relationship. They are close to universal and usually harmless unless one partner becomes upset by witnessing flirtatious activity on the part of the other or by hearing about it secondhand. Observed flirtations are especially unsettling for the possessive and insecure because they are concrete reminders that the flirter finds others sexually appealing and therefore might consider going further.

But flirtations can have positive effects too. It can be gratifying and empowering for a partner caught flirting—at a party, for example—to see his or her partner become jealous. Noticing one's partner flirting can also stimulate passion, partly because a mate's desirability is enhanced by outside confirmation, and also because mild undercurrents of uneasiness and irritation can act as aphrodisiacs.

Dalliances—casual, typically once-only encounters—are the most common explicit breaches of monogamy. Luckily, the vast majority never come to light. I believe that occasional dalliances are an aspect of erotic life better left unrevealed. As long as strict safe sex guidelines are followed, dalliances typically pose no more threat to a relationship than masturbation, fantasizing about other people, or enjoying erotic videos or stories. But should the partner learn of a dalliance, he or she probably won't take it lightly. On the contrary, even a meaningless tryst with a stranger thousands of miles from home has been known to create a crisis of trust. Dalliances are clearly not risk-free, nor is it my intention to advocate them.

Affairs are far more serious breaches of monogamy because they involve repeated forbidden meetings and often include more than just sex. Most affairs run their course quickly but others persist for years, sometimes developing into stable secondary relationships. Whereas flirtations and dalliances do not necessarily result from problems in a relationship, ongoing affairs are much more likely to be indicative of trouble.

Affairs occur under so many circumstances and for so many reasons that it's difficult to generalize about them. Studies disagree substantially about the prevalence of affairs, with reported frequencies ranging from 20 to 50 percent of married couples to well over three-quarters of long-term gay male couples. Blumstein and Schwartz mention a number of impressions[6] that closely match my observations:

- Men are more likely to have affairs than women and, if they do, to have more of them.
- When men have affairs they're usually looking for sexual variety, including opportunities to engage in sexual behaviors they don't engage in with their partners.
- Women who have affairs, especially lesbians, typically develop emotional attachments.
- Older women are much less likely than younger ones to have had affairs.
- When one partner feels less committed to a relationship than the other, he or she is more likely to have an affair.

- Neither religious beliefs nor one's degree of participation in a church or synagogue has any effect on the likelihood of an affair.

Most affairs, even ongoing ones, are never discovered or verified. Unless partners have mutually agreed to allow outside sex under clearly defined conditions, when affairs do come to light they invariably provoke a crisis, sometimes with such a profound sense of betrayal that the partnership is unable to survive it. This isn't surprising given the fact that affairs are most likely to occur when problems already exist in a relationship.

Crises, including those produced by the discovery of an affair, are never pleasant, but neither are they always bad. Some couples use the ordeal of recovering from an affair to reaffirm their commitment. They express feelings and needs they have kept to themselves for years. They also tackle tough questions about where they want the relationship to go. Some begin dating each other again, assuming nothing, as if starting from scratch.

Many believe that any dalliance or affair makes a mockery of monogamy. This notion is particularly prevalent in traditional heterosexual marriages in which monogamy is a given and not open to discussion. More than a few spouses passively agree to monogamy because they have no real choice. Many others who voluntarily place a high value on sexual fidelity don't always behave in accordance with their ideals. In my view, imperfect behavior shouldn't necessarily invalidate good intentions. When someone claims to be "mostly monogamous," he or she may be expressing a more genuine commitment to monogamy than someone who passively capitulates to its demands.

Beyond monogamy

Millions of couples retain an allegiance to the ideal of monogamy even though they occasionally violate it. A smaller number openly reject the constraints of monogamy, forging agreements that allow for some degree of sexual or even emotional involvement outside the primary bond. In the freewheeling 1970s open marriage was widely discussed. Ultimately, most couples—even those who saw value in the concept—were unable to handle the inevitable jealousies and insecurities.

In the 1980s, as concerns about AIDS and other sexually transmitted diseases increased, nonmonogamy slipped further into disfavor. Nonetheless, some couples still find that nonmonogamy makes sense for a variety of reasons, including:

- Two partners have a wide discrepancy in their desired frequency of sex.
- One partner has specific sexual desires that can't be met by the primary partner.
- One or both partners feels bored and wants outside stimulation.
- Circumstance requires partners to be apart for extended periods.
- One or both partners is bisexual.

Hardly any couples who venture beyond monogamy adopt an anything-goes attitude. Some work out tacit understandings without much talk. For example, discreet dalliances are acceptable when either partner is out of town. I've seen situations in which one or both partners pursued not altogether secret affairs for extended periods virtually without discussion, and with few obvious problems.

Other couples engage in detailed, often stormy negotiations about outside sex. Sometimes a "night off" is established when both can do whatever they please as long as they're home by a specified time. Some prefer not to hear details of outside sex while others want to know everything so they don't have to wonder. Still others only want to be informed of an emotional involvement. Virtually all nonmonogamous couples—like monogamous ones—need comfort rather than ridicule when jealousies and insecurities arise, as well as reassurance that their relationship is still primary.

As we have seen, most but by no means all gay male couples allow at least some outside sex, especially as an antidote for sexual boredom after years of monogamy. Although lots of gay men have just as much trouble making this transition as their lesbian and straight counterparts, two factors make it a little easier. In the gay male subculture there is social support for open relationships and role models for making them work.

Another crucial factor among gay men is what Blumstein and Schwartz call the "trick mentality," in which sex as recreation is separated from emotional involvement.[7] Gay male couples who want nothing more than casual outside sex usually find the adjustment relatively painless, particularly if they are sensitive to each other's feelings and continue to enjoy sex, at least occasionally, together. But many male couples face the same discrepancy as heterosexual couples: one partner seeks casual outside sex while the other can't imagine avoiding deeper involvement. For them nonmonogamy is just as stormy—and often unworkable—as it is for most opposite-sex or lesbian couples.

MASTURBATION: HONORING THE PRIVATE SIDE OF SEX

Despite a widespread belief that good sex with a partner renders private sexual acts superfluous, many partnered men and women continue enjoying masturbation and fantasy. Those who focus their sexual desires exclusively on their partners are most likely to be troubled by the realization that certain aspects of their partner's sex life have little or nothing to do with them. A few go so far as to react with the kind of bitter jealousy most of us reserve for sexual indiscretions with other people.

Not only is long-term loving not diminished by solo sex, but some people discover that an active private sex life is an essential ingredient for erotic vitality. On the most practical level solo sex keeps people in touch with their bodies. Virtually all sex therapists agree that conscious self-stimulation is one of the most effective ways for a woman to learn how to have orgasms reliably. Similarly, men with a tendency to ejaculate rapidly can use deep breathing and muscle relaxation during masturbation to increase awareness and control of their responses, lessons readily transferable to encounters with a partner.

On a less clinical level, the erotic fantasies that typically accompany masturbation offer direct access to one's CET, the inner wellspring of passion. Even when heightened desire is focused on a fantasy partner, in many instances the actual relationship benefits.

Solo sex has yet another major advantage that is too often ignored. It is without question the most readily available way for lovers to cope with small to moderate discrepancies in desired sexual frequency. I've seen dozens of marriages that have literally been saved by masturbation. Rather than pestering a reluctant partner, the one who wants more sex satisfies some of his or her needs alone. Arguments are reduced, which often increases the quality of the sex they do have together. This approach doesn't work so well when disparities in sexual desire are extreme because then the person with higher desires has to depend primarily on masturbation to satisfy his or her needs. Neither is this method effective for someone who derives little or no enjoyment from masturbation.

Secrecy versus privacy

To understand the role of masturbation and fantasy in committed relationships it is crucial to make a clear distinction between *secrecy,* hiding significant information, and *privacy,* the right to maintain a nonrelational sphere of existence. Secrecy hurts intimate relationships; privacy enhances them. Unfortunately, many people don't recognize the difference. They fear secrets, and therefore they resent privacy. Ironically, when they refuse to recognize legitimate privacy rights, instead of making themselves more secure, they create the very secrecy they fear. Erotic couples honor each other's sexual privacy. In most cases this comes down to not discussing masturbatory activities so that the private sexual realm never becomes an issue.

Sometimes, however, one partner unwittingly stumbles upon information that is genuinely perplexing or upsetting. I've worked with couples of all sexual orientations in which one person discovered a stash of erotic magazines or videos and became distraught by the idea of his or her partner fantasizing about someone else. Some people are mistakenly convinced that masturbation itself is a sign of trouble in their relationships.

Believe it or not, I've also known more than a few couples

who argued about vibrators, particularly their amazing ability to produce multiple orgasms on demand. Many women find the intense vibrations, enhanced by fantasy, to be quite compelling. But some of the partners of these women, male or female, see the machine as threatening because its feats of extended stimulation cannot be duplicated by mere mortals.

Carl and Sandra: The plug-in rival

Carl didn't pay much attention when Sandra brought home a vibrator until it became a permanent fixture on her nightstand. A showdown came one morning when Sandra asked Carl if she could use the vibrator on her clitoris while he was inside her. He pulled out immediately and turned away. Later he accused her of preferring sex with "that damn machine." Sandra tried to explain that she needed a little extra stimulation to climax during intercourse and that the vibrator was helping her to become more easily orgasmic. "I'd still be making love with *you*," she assured him.

Sandra's difficulty with orgasms during intercourse, along with Carl's inability to have intercourse for more than about ten minutes without ejaculating, had brought them to therapy. Logically, you might expect Carl to support any method that could help Sandra reach orgasm. But like many men Carl felt less than adequate because he couldn't *make* her come before he did. He made his problem much worse by labeling himself a "premature ejaculator."[8]

The pressure on Sandra was unbearable because she hardly ever climaxed during intercourse, no matter how long it lasted. It was a major breakthrough when Carl agreed that Sandra could rub her clitoris as they made love. Precisely because this seemed to be working so well, Sandra thought Carl might accept the vibrator. But because he was so jealous, Sandra put away the vibrator although she continued to enjoy it privately. The subject never came up again until, having made great strides in their communication and comfort with each other, they were almost ready to stop therapy.

Much to Sandra's surprise, one Friday night Carl whispered

just before they drifted off to sleep, "Why don't you get out the vibrator. Maybe we can use it in the morning." They both woke up in a sexy mood. Carl requested instructions on vibrator technique, then devised a position in which he could enter her and simultaneously hold the vibrator to her clitoris. She, of course, climaxed explosively, for which she almost felt the need to apologize. But apparently Carl had resolved his rivalry with the vibrator. From that point on a muffled buzzing sound could often be heard in their bedroom.

Most people require a minimum of paraphernalia during solo sex—perhaps a little lubricant, a favorite piece of erotica, and a towel or tissues for cleanup. But some masturbatory props can, if discovered, be even more upsetting than Sandra's vibrator was to Carl.

Melody and Herb: Herb's secret

Less than five minutes after she and her husband, Herb, entered my office, Melody was telling her story between sobs. She had arrived home early from an evening meeting and, as always, was looking forward to seeing Herb. She decided to surprise him, so she tiptoed toward the bedroom. There she encountered Herb wearing a lacy nightie he had recently bought for her, feverishly masturbating. What followed might have been comical except that Herb's secret—that wearing feminine lingerie was a turn-on—threw their tranquil relationship into a crisis.

Herb explained that he began eroticizing women's undergarments as a teenager. He masturbated in the bathroom where his mother and older sister hung their frilly things to dry. At first just the sight of these uniquely female items excited him, but soon the satiny textures captivated him as well. It was exciting to rub against them or put them on. Although you might assume that there was an incestuous element involved, to Herb these were not Mom's or Sis's items. They were symbols of his love of femininity. It wasn't that he felt feminine, but rather he had a desire to be close to the feminine, to revel in its mysteries.

Although traditionally feminine in appearance, Melody had little use for frilly garments and often wondered why Herb insisted on giving them to her. Not only did she feel betrayed, she also feared that this secret might be the "tip of the iceberg." Was he having kinky affairs? Was he gay? Was he a peeping Tom? The man she thought she had known so intimately suddenly seemed a stranger. Luckily, Herb was eager to explain how he developed his unusual interest, how he had often wanted to tell her about it, and how he hoped she might one day come to bed wearing the sexy items he had given her. Nothing was different, he reassured her, except that now she knew that he "masturbated with props."

Herb expressed the hope that they could use this crisis to talk more openly about sex and become more experimental, maybe even acting out some of their fantasies. Melody, on the other hand, insisted that she didn't fantasize, didn't want to be adventuresome in bed, definitely didn't want to be given any more lingerie, and hoped that Herb could somehow get over his fetish. But more than anything she made it abundantly clear that never again did she want to be confronted with this private aspect of Herb's sexuality. If he couldn't "get over it," he at least had to keep it to himself. Reluctantly, Herb agreed. With the wall of privacy reestablished they resumed their warm and loving relationship and left therapy.

Sharing the private world

Some couples, unlike Melody and Herb, discover that they thoroughly enjoy sharing each other's private sex lives, either directly or indirectly. For these adventurous pairs, masturbating together is a provocative way for each to demonstrate how he or she likes to be touched. What better way to learn precisely what works for your partner? Once you are past the initial awkwardness of participating in such a private act, the instruction process can be quite titillating.

Sharing erotic fantasies is more complicated. The close link between masturbation and fantasy is one of the key reasons that solo sex is usually kept private. Many people find it difficult to masturbate without fantasizing, and difficult to fantasize freely in the presence of their partners. And even though it is known that men and women frequently fantasize about other people while making love with their partners, how many of us really want to hear about it? Even more challenging is the fact that fantasies are powerfully shaped by the fantasizer's CET. And although every sexual relationship is influenced by the interplay of both partners' CETs, openly discussing them is an entirely different matter.

ACCOMMODATING EACH OTHER'S CET

One of the most unsettling of love's rude awakenings is the realization that your partner's deepest erotic yearnings are unexpectedly at odds with yours. I'm not referring to the everyday incompatibilities of timing, frequency, rhythm, and styles of touch that every couple encounters as their sex lives mature. There are much more difficult situations in which a person's CET is unable to find an avenue of expression within the relationship. These serious erotic incompatibilities frequently don't come to light until the relationship is well established.

Theresa and Rita: The politics of fantasy

Despite the protests of her lover, Rita, Theresa requested that she meet with me alone. They were already in couple's therapy with a woman therapist, working out disagreements about parenting Rita's daughter from a previous marriage. They also had a variety of other concerns, among them an unwelcome phenomenon known as "lesbian bed death."[9] What little sex they still had was decreasing rapidly after less then two years together. Heated discussions seemed to be getting them nowhere, although they obviously loved each other and had a marvelous closeness and a community of friends.

Theresa had recently confided in Rita that fantasies of dominance and submission had always turned her on and how disappointed she was that there appeared to be no room for playing with power in their sexual relationship. She had fully expected Rita to be upset because Rita felt so strongly about equality in all things, including sex. Theresa explained, "I tried to be as diplomatic as possible when I told her I long for the sexy energy when I submit to her or feel her surrender to me. But I wouldn't *dare* tell her that I dream of commanding her to get on her knees and lick my pussy, or how much I would love it if she pinned me against the bed with the full weight of her body and called me a slut."

"Why not?" I asked as if I weren't aware of the controversy dividing lesbians about the relationship between sex and power. "Because dominance and submission is based on heterosexist models that demean women and cheapen sex," she responded with the most amazing mixture of sarcasm and heartfelt conviction.

She was unwilling to discuss this with a woman therapist, especially a lesbian, for fear that she would be scolded. Like her community, she was conflicted and doubted she could risk telling *any* lesbian about what really turned her on. And because this was the early 1980s, few articulate lesbian voices had yet emerged to call for sexuality to be released from the choking demands of political correctness. Fortunately, the situation has since changed.

Although she was outspoken, tough, and assertive, Theresa moved and spoke with feminine grace. One of her complaints about the lesbian community was its insistence that everyone be androgynous—nobody too "butch" or too "femme." Theresa liked being an outspoken femme and was naturally drawn to butch women like Rita. She found the contrast a turn-on despite worries she was "mimicking heterosexual roles."[10]

Lesbians continue to debate the acceptable parameters for sexual fantasy and play, especially within the framework of loving relationships. This question arises for men and women of all sexual orientations: Because so many people are harmed by large and small abuses of power, and because sexualized abuses of power are especially demeaning, is there any room in relationships of mutual respect and caring for sexual power play? The question is particularly difficult for anyone whose fantasies include images of humiliation, or who has been sexually abused.

Seth and Lenore: Was it rape?

About the same time Theresa and Rita were grappling with their problems I began working with Seth and Lenore, who were facing a major crisis after nine years of marriage. One day Seth had become amorous with Lenore in the kitchen and before long he entered her. They were standing up with Lenore pressed against the kitchen counter. As their excitement escalated, Seth grew increasingly aggressive. Lenore was enjoying herself until she noticed Seth's glazed eyes and the contorted expression she later called the "look of a rapist." She was especially frightened because she had recently been thinking about an incident of date rape from her high school days—long before anybody openly recognized the phenomenon.

After Seth ejaculated and as his attention returned to Lenore, he saw a look of horror and rage in her eyes. Then she pushed him away and ran off sobbing. Seth confirmed for me in one of our individual meetings that at the height of his excitement Lenore had indeed temporarily become purely a sexual object, and he had been thrilled by the feeling of "fucking her mercilessly." His internal imagery, however, was not of rape but of unbridled lust and a receptive woman overwhelmed by the power of his sexual energy.

Lenore referred to the event as "the rape." Although Seth was shocked by this characterization, he felt too guilty to protest. For Lenore, the fact that at the height of his excitement Seth often became less tender and more animalistic felt perilously similar to her memories of the date rape.

Seth had enjoyed pornography since adolescence, so he began reading about it, especially the notion that all porn degrades women by objectifying them. He decided that never again would he abandon himself to lust while making love with Lenore. The topic receded into the background as they emphasized warm, intimate sex, which they both enjoyed. They also worked on deepening the trust between them. They left therapy feeling good about their relationship. But during one of our individual wrap-up sessions Seth told me he missed the heights of excitement he used to reach before he decided to restrain himself. Although he firmly believed this was a small price to pay for a better relationship with Lenore, he nonetheless mourned the loss.

Theresa and Rita moved in the opposite direction. Rita decided she was willing to experiment with power fantasies once she realized how important it was to Theresa. Luckily, there were no experiences of sexual abuse in Rita's background, so only her political objections made her reticent. Rita, a very self-aware person, recognized that she too could genuinely enjoy mild power play with Theresa.

Rita even hoped that Theresa might be willing to participate in one of Rita's favorite fantasies, although she held serious doubts about the wisdom of acting out in erotic role playing a scenario that would horrify her in real life. She finally decided to tell Theresa about the fantasy that always had the power to excite her. In it she's a svelte, sought-after stripper working in a tawdry nightclub overcrowded with gawking men. Out of the throng a beautiful woman appears, takes her hand, and leads her to a secret back room. There she lies on white satin sheets, aching for this strange women to worship her body, to kiss her everywhere with unbearable sensuality.

Luckily, Rita and Theresa took readily to the imaginative joys of erotic role playing and shared a similar style: rather than acting out every detail explicitly, they preferred to focus on simple symbolic acts as fantasy stimuli. It was obvious to both that their newfound freedom dramatically improved their sex life.

Two similar dilemmas; two radically different outcomes. In Seth's relationship his CET was excluded, resulting in a welcome improvement in tenderness and intimacy. The lesbian couple, in contrast, overcame pervasive political pressures and exploited "incorrect" imagery for fun. Somewhere in between are countless other couples who never overtly discuss their CETs yet become reasonably adept at sensing the private turn-ons of their partners and providing complementary movements, postures, and words to stimulate the lover, without openly acknowledging what they're doing.

THE DANCE OF INTIMACY AND PASSION

All erotic couples share a special skill: they are able to adapt gracefully to new challenges. The more I study long-term couples, the more impressed I am by their versatility and creativity. Often these couples surprise themselves with the solutions they discover through trial and error.

When newer couples hear about the compromises and adaptations made by more established ones, they may feel discouraged. They don't yet understand that while countless good relationships exist, ideal ones are figments of the imagination. An example might be a couple who, after years of struggling with incompatible living habits, decide to live separately while actively maintaining their connection. Once they are freed of constant petty conflicts their intimacy improves.

One of the most widespread and destructive myths is that healthy couples have a consistent sex life. Human sexuality didn't evolve to be expressed with clockwork regularity, a fact that becomes increasingly obvious with the passage of time. Rare are couples who don't experience dry spells, especially in today's two-career, high-stress households. The way a couple responds to these inevitable fluctuations has a greater effect on the long-term viability of their relationship than the dry spells themselves.

Some couples reach a point when they mostly or totally stop having sex. Although nonsexual couples are less likely to feel satisfied with their relationship than couples who continue to have sex, some are relatively content. The degree of importance given to sex varies tremendously, particularly among older couples. Many continue to enjoy active sex lives, some more than ever because they have extra time together and no worries about pregnancy. Others evolve into nonsexual companions.

You might find it distressing that there are so many nonidyllic relationship options. But there's another way of looking at it. The many faces of couplehood can be inspiring, for they demonstrate the adaptability and endurance of long-term connections, even in an age when commitments so often seem transitory. As partners free themselves from the tyranny of one predetermined style, love finds more ways to flourish.

If both intimacy and passion are to thrive beyond the limerent period, there must be a recognition that the two spring from separate and distinct motives. Intimacy is engendered by the desire to know every detail of the other's dreams and fears. Passion, however, is felt when one gazes at the beloved from a distance and appreciates him or her as an individual who can never be fully known.

10

SIGNPOSTS TO EROTIC HEALTH

Evaluate your sexual well-being from the paradoxical perspective.

While exploring the many dimensions of eroticism, you've undoubtedly wished, as I have, that it could be simpler, more predictable. But in the realm of eros, where expectations are so frequently shattered, where obstacles and emotions can either enhance or inhibit passion, and where almost everything is more complex than it first appears, drawing definitive conclusions from what we've learned is anything but easy. This lack of certainty is especially apparent when we try to describe key characteristics of healthy eroticism—our goal in this chapter.

Evaluating health from the familiar pathology perspective is relatively straightforward. If a doctor doesn't diagnose an illness, you're considered healthy. Making the proper diagnosis might be tricky, but the concept is reassuringly uncomplicated. Following a similar line of thinking, psychologists and sex therapists tend to conceptualize sexual health as the absence or reduction of distressing symptoms such as sexual dysfunctions, inhibited desire, incompatibilities between partners, or compulsive urges to reenact unfulfilling or harmful turn-ons. With rare exceptions, clients who make progress in resolving symptoms like these do indeed feel happier and healthier. It is erroneous, however, to assume they all would have been better off without their symptoms in the first place.

A serious limitation of the pathology perspective is its one-dimensional view of sexual troubles. Although I've never met anyone who was happy to have a sex problem, I've worked with dozens whose unwelcome symptoms turned out to be opportunities for growth. Not only did their problems push them to confront patterns that were depriving them of satisfaction, but as they searched for better alternatives they found new levels of self-respect and confidence. Time and again I've learned from my clients that it's not so much whether they have problems but how inventively they deal with them that most determines their degree of wellness.

According to the paradoxical perspective, grappling with life's dilemmas is a central and unavoidable aspect of the erotic adventure. Once we grasp this truth, we can't possibly visualize healthy eroticism as a fixed, problem-free state—nor should this be our ideal. Erotic well-being expands as we acknowledge and integrate contradictory emotions and motivations within ourselves, while also learning to cope with them in others. Those on the path to erotic health discover that problems and potentials are two aspects of a whole.

Unfortunately, there's no simple way to gauge how effectively a person is rising to these challenges. Nonetheless, implicit in the paradoxical perspective are vital indicators—I call them signposts—many of which are already familiar to you. The best way to evaluate where you are on the road to wellness is to review some of the noteworthy abilities you've been actively cultivating while reading this book:

- Enjoying peak experiences while also using them as avenues to self-awareness (Chapter 1)
- Recognizing and accepting the role of emotions—including "negative" ones—in your turn-ons (Chapter 4)
- Identifying childhood challenges and psychic wounds that fuel your strongest passions (Chapters 3 and 5)
- Acknowledging when ingrained sexual scripts are working against you (Chapters 6 and 7)

- Knowing when and how to commit to necessary, self-affirming changes (Chapter 8)
- Building an "interactive zone" between love and lust (Chapters 6 and 9)

Only someone who knows you extremely well can assess how comfortable you are with these abilities, which is why I suggest you do it yourself. As long as you don't take this exercise too seriously, try rating yourself either *low, medium,* or *high* for each signpost. *Low* means you haven't yet thought much about that skill or else you're having troubling making it work for you. Use a *medium* rating wherever you see progress, even if it's sporadic or tentative. *High* is for signposts that have become regular—though imperfect—aspects of your eroticism.

No scoring system can possibly capture the subtle manifestations of such complex indicators. Even so, your ratings do contain hints about where your erotic health is thriving and where it may be limited or stalled. Focus first on your highest ratings because they're inspiring; they show what you're capable of. Because of the mundane connotations of medium ratings, you might assume your progress on these signposts is merely average and of minimal importance. In actuality, medium-rated signposts are likely to be on the cutting edge of your erotic development. Here you can often find an inner struggle, perhaps quite subtle, between a genuine desire to grow and an urge to play it safe. Clarifying that struggle is a major step.

It's tempting to look down at a low rating but try not to. This isn't a test, and your ratings aren't grades. Besides, low ratings are as informative as any other kind. Ask yourself: "Does this signpost make any sense to me? If so, what's preventing me from working with it? Am I indifferent? Or could I be avoiding or rejecting something about it?" Unless you're simply indifferent, a low rating may be a sign that you're holding back and is therefore a quiet call to action. If you can identify just one small step to ease you forward, so much the better.

If you find yourself formulating signs of health more meaningful than any I have suggested so far, honor them. Health indicators that come from firsthand experience are especially likely to affect your actions. Write about them or talk to an intimate friend or lover. Of course, not every idea stands up to scrutiny. But with rare exceptions, those who make an effort to define what healthy eroticism means to them ultimately become its promoters.

There's no better way to become engaged with the great issues of erotic health than to review the highlights of your voyage of self-discovery. Now that you have launched that review, the time has come to add several other signposts to your list, ones we've touched on only briefly or indirectly before. Yet I'm sure you'll sense interconnections with the fundamental themes of your journey:

- Clarifying your personal values, including:
  - respecting self and others
  - facing your erotic shadow
  - claiming the responsibilities of freedom
- Differentiating fantasy from action
- Nurturing children's sexuality
- Appreciating erotic diversity

The fact that I've waited until now to call your attention to these signposts in no way diminishes their significance. On the contrary, my choice of timing comes from years of observing when growing people naturally tend to bring these aspects of eroticism most productively to the fore. There's no doubt these issues are relevant at all levels of erotic development—and are thus basic. Yet they acquire a far deeper significance for those who have first investigated their eroticism. With greater self-awareness comes an increasing need to grapple with these crucial questions, and to define where we stand. The dual quality of these signposts—both basic and advanced—means that anyone can benefit from considering them. Those who have already dedicated themselves to conscious eroticism, however, are prepared to gain the most.

CLARIFYING PERSONAL VALUES

Throughout this book I've encouraged you to set aside judgments—including moral ones—so that you might perceive more accurately the breadth of the erotic experience. To the extent you've been able to do this, you have undoubtedly encountered crucial truths that only accepting eyes can see. But as you discover the benefits of examining eroticism with an open mind, what becomes of your moral beliefs and values?

Some people confuse a conscious suspension of criticisms with an amoral, anything-goes attitude toward sex. However, the vast of majority of people I work with have a very different response. They find that a courageous, nonjudgmental exploration of their eroticism ultimately prompts them to think more deeply about how to define and promote what might be called *right sexual conduct*. The goal of right conduct is to bring one's behavior into alignment with one's values—not perfectly, of course, but as closely as possible. As you've probed the inner workings of your erotic mind and cultivated a greater respect for its irrational power, you've built a strong foundation upon which to explore and commit to an ethical approach to erotic life.

When moral questions arise in the context of active self-examination, you may be dismayed to discover—as most of my clients have been—that your early moral training was so fundamentally antierotic that it's of limited use to you now. The sad truth is that conventional morality asks neither that we understand nor that we respect our eroticism. Instead it insists that we treat certain sexual desires with mistrust while blindly adhering to rigid codes of conduct. Most of the time these codes focus on prohibited acts such as nonreproductive sex, sex outside marriage, homosexuality, masturbation, oral or anal sex, and so on. Because it fails to come to terms with the multidimensional nature of eros and invests so much energy into fearing and suppressing its disturbing or potentially destructive aspects, conventional morality often works against healthy eroticism rather than fostering it.

Another problem with traditional morality is that it usually requires allegiance to some strong authority such as a dominant religion or widely accepted cultural norms. In pluralistic societies where there is no single authority, "thou shalt nots" lose some of their potency as once sacrosanct beliefs about proper conduct are scrutinized and challenged. Doctrines that reduce profound and complex ethical issues to simplistic moralisms tend to be particularly brittle in the face of such scrutiny, partly due to their inflexibility, and partly because they are so often founded on little more than ignorance, custom, or prejudice. Consequently, when externally imposed moral codes weaken, those who once believed in them often feel bewildered and disillusioned. More than a few embark on a frantic and fruitless search for new absolutes.

As you have looked at the contradictions within the erotic mind, you've become familiar with another feature of don't-do-it morality: prohibitions have of a way of increasing one's fascination with the very acts they seek to suppress. You know that children who grow up surrounded by sexual restrictions stand a much better chance than most of becoming adults who are ambivalently and guiltily drawn to forbidden behavior.

These and other flaws inherent in conventional morality cause some people to swing to the other extreme and reject *all* ethical considerations as antierotic. Some even convince themselves that sex and morality have little or nothing to do with each other. For a time disconnecting the two can be liberating, as it was for so many during the sexual revolution of the 1970s when a popular slogan was "If it feels good, do it!" Before long, however, that credo proved to be as vacuous as the pious moralisms it sought to supplant. Unless it is grounded in a conscious understanding of how one wishes to live and what truly matters, sexual liberation ultimately becomes an empty goal.

Fortunately, there's an alternative approach that avoids the pitfalls of either extreme: *The erotically healthy person develops a clear set of ethical values that possess intrinsic personal meaning and applies them in the sexual arena.*

Personal values frequently overlap with traditional morality, but they operate quite differently. When we obey or pay lip service to standards of conduct passed down from on high, we tend to do so out of obligation, fear, guilt, or habit. Conversely, we honor our own values because we genuinely believe they have a direct impact on the quality of our lives. Value-based decisions are more likely to shape how we actually behave when they are forged from firsthand experiences and careful observations.

I'm not saying we should ignore the collective wisdom that

can readily be found in all moral traditions. But I do think the value of that wisdom increases dramatically when we test it in the laboratories of own lives and, when appropriate, choose to make it our own. I'm also convinced that ethical values are far more likely to affect our actions when we focus on fundamental principles rather than complicated lists of rules that can easily distract us. Three such principles are of utmost importance to everyone's erotic well-being: (1) respecting self and others, (2) acknowledging rather than denying one's dark side, and (3) claiming the responsibilities of freedom.

RESPECTING SELF AND OTHERS

We have observed how, with rare exceptions, an awareness of otherness is essential for both passion and emotional closeness. Relatively few people are interested in mere reflections of themselves. The vast majority of us are keenly aware that desired ones are attractive because they exist outside the physical and psychic boundaries that define where we end and they begin.

We know that lusty attractions tend to objectify the desired one to a certain extent, especially during moments of high arousal. Contrary to what some people believe, lusty objectification can be thoroughly exciting and validating for everyone involved. Positive objectification accentuates the separateness of the other by using one or more of the other's features as focal points for passionate energy. Problems occur when someone is seen *only* as an object of desire without any recognition of his or her feelings, preferences, or rights—in short, when basic respect is lacking.

The desire to reach out for connection springs from a similar realization of separateness, for only two distinct individuals can share intimacy. When a compelling romantic response generates an intense urge to merge, most of us never completely forget that the self and the other remain two separate entities, even as we revel in feelings of oneness. When partners do lose sight of each other's individuality, with one treating the other merely as an appendage, the relationship ultimately sours and both feel depleted rather than enriched by the exchange.

In erotic life, recognition and respect for the separateness of the other is manifested, first and foremost, in a profoundly simple conviction: *sexual contact is appropriate only when both partners give their full consent.* This means never willingly submitting to unwanted sex no matter how persistent the pursuer, and never pressing another past the point of an unambiguous "no." It means remembering that no one ever owes sex to anyone. And it certainly means that those who become so desperately "lovesick" or resentful that they resort to pestering, stalking, or coercion must seek help—or be stopped by the community. Healthy eroticism simply cannot exist unless requests to be left alone are honored.

Even when the importance of giving and receiving consent is recognized, ambiguities still occur. Some lovers are ambivalent, simultaneously wanting and not wanting sex. Others enjoy being seduced while feigning reticence. Still others prefer *not* to give explicit verbal consent—when, for example, they want their partners to take total control—requiring the participants to interpret subtle nonverbal cues.

There are also ambiguities about who can legally or psychologically give consent. The age of consent is set by law, even though some young people are obviously better equipped than some adults to know and state clearly what they want. Many factors undermine one's capacity for giving consent, including severe emotional distress, intellectual impairment, or the excessive use of alcohol and other drugs. Inequities of power also complicate consent, often enormously.

There will always be attempts to remove the ambiguities of consent through laws and regulations, some of which are, of course, essential.[1] But in the final analysis only a heartfelt respect for ourselves and others can guide us through eros's ambiguities—precisely *because* we recognize their inevitability. Self-respect requires that we speak up when sexual advances are unwanted, while vigorously resisting any temptation to perceive ourselves as helpless victims. To respect the other means simply that we listen to his or her wishes and, if we are uncertain about what they are, to ask for clarification. There are also occasions, which are difficult to evaluate, when respecting others requires us to figure out a way to intervene if we observe someone being mistreated who seems unable to prevent it themselves.

FACING YOUR EROTIC SHADOW

If we always knew exactly what was ethically right and emotionally healthy, we wouldn't need to think about values. Our actions and principles would coexist in perpetual—and undoubtedly boring—harmony. But this isn't the case, certainly not with eroticism. Even the most virtuous among us occasionally experiences urges to violate his or her own values, no matter how freely chosen or enthusiastically embraced. Our explorations have shown that both love and lust have dark sides, which all who care about sexual health must confront head-on.

People who face their own imperfections tend to be more conscious of the implications of their actions than those who deny that they have an erotic shadow or pretend their motives are always pure. Unless we are aware of our capabilities—for good and ill—even the most well-intentioned values become detached from the rough-and-tumble of real life and slip into irrelevancy. It is extremely beneficial to take an unflinching look at when, how, and why you have resorted to manipulation, deceit, or other predatory practices in the pursuit of sexual gratification, no matter how subtle you may have been about it. The purpose of these acknowledgments is not confessional. And self-flagellation only makes matters much worse. The aim is to confront your shadow impulses consciously and use your discoveries as sources of self-knowledge.

CLAIMING THE RESPONSIBILITIES OF FREEDOM

To discover your erotic potential you must be free to choose what is right for you. But this isn't as easy as it sounds because each choice brings you face to face with the consequences of your actions. Whether you're deciding what to do about birth control, developing a policy to avoid sexually transmitted diseases, or grappling with any of eros's practical tasks, once you've made a choice, you are responsible for the outcome. Claiming ignorance—even though one person never possesses all the facts—isn't an option.

The key is to remain conscious of your actions and their consequences without becoming self-conscious, a state antithetical to emotional and sexual well-being. Learning to balance the thoughtfulness necessary for responsible behavior with the freedom from worry necessary for good sex takes effort. No wonder some people struggle with it constantly. In the age of AIDS, when the wrong kind of spontaneity invites disaster, we're reminded in a dramatic way that smart ideas don't count until they're translated into behavior. The tension between perils and pleasures, however, is as old as humanity.

The ability to accept responsibility while enjoying freedom is a hallmark of psychological maturity. Only a reasonably well-integrated person can simultaneously pursue these two sometimes contradictory objectives. But this is what we all must do.

Erotically healthy people establish safe parameters within which to let themselves go. As important as it is to have a sensible plan for avoiding unwanted consequences from sex, also needed is the capacity to monitor one's behavior, even during heights of passion, to make certain the plan is being followed. Those who practice this skill eventually learn to monitor themselves subliminally. Then they can be fully engaged in giving and receiving pleasure unless an internal alarm signals them to stop and think. Even those who find excitement on the dangerous edge can become amazingly adept at flirting with risks while taking the necessary steps to prevent actual damage.

SELF-ASSESSMENT

The best way to clarify your values—or to become clear that you're not yet clear—is to put them into words. You might begin by making a list of key principles that mean the most to you. Sometimes it's even more beneficial to initiate a series of conversations with a trusted friend or lover. The give and take of intimate discussion has a way of pushing you to clarify how you truly feel. The goal of these talks isn't agreement but mutual understanding.

Pay special attention to discrepancies that exist between your conscious attitudes and your deep-seated beliefs. I was recently talking with a woman client about her inability to have orgasms with her boyfriend. At first she expressed a conviction that women are every bit as entitled to sexual pleasure as men. But as our discussion progressed, she recognized within herself inherited beliefs that she didn't accept consciously at all, such as the idea that she should have "saved" herself for her future husband, or that sexually unrestrained women aren't respected by men. It's important to recognize that subconscious beliefs change much more slowly than conscious attitudes. The process of aligning the two takes time if, in fact, they can ever be made to match precisely.

When it comes to erotic health, nothing is more important than the interplay of eros and ethics. The hard part is knowing—and acknowledging that you know—how well you're handling that interplay. Here is one of the best indicators: if you can usually tell in a specific situation when it's appropriate to clarify and deepen your moral convictions and when it's better to set them aside, your movement toward greater health is virtually assured. When circumstances call for understanding yourself and others, suspending judgments sharpens your vision. On the other hand, the more you understand, the more you'll count on your values to translate wisdom into action.

DIFFERENTIATING FANTASY FROM ACTION

One of the beliefs most destructive to emotional well-being is that thinking or feeling something is the same as doing it. This idea runs particularly deep in cultures shaped by Christianity. In the Bible, Jesus says, "But I say unto you that whosoever looketh on a woman to lust after her hath committed adultery with her already in his heart."[2] Jesus is making the point that morality involves more than mere adherence to the letter of a law, but he is also setting a standard that is impossible to attain.[3] Many Christian children and adults are encouraged to be painfully aware of "impure thoughts" and in some traditions to confess and pay penance for them.

It's not my intention to refute Christianity except on this one point: thoughts are *not* the same as actions. We must be clear that humans are simply not equipped to corral every wayward thought and feeling, to make them conform to our ideals, or to banish the ones that don't. We all know from experience that trying not to think about something is the quickest way to become obsessed by it. In every area of life, it is crucial to differentiate how we behave from all forms of internal experience—thinking, imagining, fantasizing, dreaming, and daydreaming. Making this distinction is particularly essential in erotic life.

Those who are most upset when their thoughts and fantasies don't match each other often become so distracted by guilt and self-criticism that they have trouble making ethical decisions in their real-life behavior—where it truly matters.

Erotically healthy people recognize that sexual fantasies and behaviors operate in two separate yet interrelated spheres. Consequently, they grant themselves greater imaginative freedom than those who are less healthy. People who function in the world effectively and with respect for others are noted not for the purity of their thoughts but for the wisdom of their choices.

At the opposite end of the spectrum are those who consider imaginative activities trivial or bearing little relationship to the rest of their lives. It is true that the majority of fantasies have little or no significance beyond the immediate stimulation they provide. But as we have seen, key fantasies—especially those based on our CETs—express crucial emotional realities that *do* affect the way we see ourselves and others and, in turn, how we behave. *Erotically healthy people not only enjoy their fantasies but also use them to gain insights into their emotions and motivations.*

Consider a reasonably informed, well-functioning woman who recognizes undercurrents of hostility and revenge in some of her favorite fantasy turn-ons. She knows that such emotions are common in the erotic landscape and therefore feels little need to berate herself for unloving thoughts. Neither does she deny them or downplay their significance, but she does her best to understand what her fantasies reveal about unresolved emotional conflicts from earlier in her life. She also examines how her fantasies may influence her selection and treatment of actual partners.

Another important sign of a healthful relationship with one's sexual thoughts is the ability to claim a widening sphere of choice, not so much about whether to have a fantasy but rather how and when to give it free rein. Therefore, a man might choose to set aside a titillating image of being serviced by a harem of buxom blonds so he can concentrate on loving feelings toward his wife, even though he realizes that his wife's body gives him less of a charge than the bodies of the young blonds.

When healthy people exercise choice in their fantasy lives, their methods are subtle. Influencing the flow of fantasy is only possible for those who don't strive for total control. The ability to turn one's attention temporarily away from a fantasy is fostered by the realization that one can always come back to it later. The man with the buxom-blond fantasy told me he most enjoyed affectionate sex with his wife when he allowed images of the blonds to come and go freely without struggling with or worrying about them. His goal wasn't to banish the blonds but simply to bring his attention gently back to his wife.

The distinction between fantasy and action is particularly crucial when it comes to the dark impulses so prevalent in some of our erotic fantasies. Dr. Stoller makes an important point:

> Most people do not get excited doing with real people the cruelties they most want but only imagine. That may be the difference between those who are perverse and those who are not: the perverse, when really doing what they think about, can still remain excited.[4]

I would add, however, that acting out a seemingly humiliating or degrading scenario with a real person can be a consensual act of shared imagination, fundamentally different from hurting another through manipulation or coercion. The key distinction is that you are *pretending*.

DANGER IN THE ANTIPORNOGRAPHY MOVEMENT

Because women are disproportionately harmed by sexual assaults, it is completely understandable that they demand action against them. But I respectfully suggest that antiporn feminists are following a course that will ultimately make the situation worse rather than better.

If a brilliant misogynist were to devise a plan to increase the sexual mistreatment of women, he surely would begin with a propaganda campaign to convince everyone that fantasy and behavior are interchangeable. If it worked, we would gradually come to believe, as many already do, that demeaning, disgusting, or brutal images are the same as destructive acts. Unfortunately, this is the same dangerous message the antipornography movement is unwittingly propagating. Those who target fantasy representations for suppression are subtly blurring the boundaries between fantasy and behavior.

Lack of clarity about the difference between fantasy and action is responsible for more destructive sexual acts than all the violent pornography in the world—of which there is relatively little, especially compared to the unrelenting depictions of violence in the mainstream media committed by and against sexy people. In my limited experience working with sex offenders, I've consistently seen their difficulties distinguishing thoughts from acts, a curious phenomenon also noted by professionals who specialize in his area. In addition, some offenders appear to have stunted imaginative abilities; they can only act out what millions of others explore within the safe confines of fantasy.

In spite of my opposition to suppressing porn, I disagree with those who claim that pornography is totally benign. I've seen how porn sometimes contributes to very real problems. For example, adolescent males who base their sexuality primarily on porn usually find themselves tragically ill-equipped for real relationships. In addition, dozens of men have consulted me for help with sexual dysfunctions that were at least partly traceable to the unrealistic exaggerations of porn. We also know that some men prone to commit sex crimes stoke themselves up with porn. As they "space out," they detach themselves from reality and reduce what may already be a limited capacity to perceive their victims as more than objects in a fantasy.

I've also seen the benefits of porn. Sometimes it helps restimulate desire in those whose eroticism has gone into hibernation. For others it facilitates the ability to fantasize, thus boosting arousal and contributing to an overall sense of sexiness. Still others use images gleaned from porn to help them become orgasmic. With the increasing availability of erotic videos, many couples enjoy their stimulating effects together. Most commonly of all, porn is a harmless accompaniment to masturbation.

Porn is essentially fantasy and fantasies can be beneficial, destructive, or inocuous. Like most aspects of eroticism, porn is too complex in its meanings and uses to adopt one single attitude toward it. Of course, much of it is deliberately designed to offend traditional morality. A small segment of it is truly horrifying because it gives expression to the darkest impulses of the human psyche. Nonetheless, perceiving porn as the enemy and seeking to eradicate it (a losing proposition) is counterproductive because it diverts our attention from the more realistic goal of forging a societal consensus about acceptable behavior.

Although I think the repressive, censoring impulse does a dangerous disservice to the fight against sexual mistreatment, I consider the *expressive* impulse—the growing willingness of women to speak forcefully about their experiences with male sexuality—to be exactly what's needed. As women break their long silence they make all of us face the predatory expressions of the erotic impulse. Following the lead of courageous women, some men are now also revealing how they too have been hurt by sexual exploitation.

Most important, greater expressiveness helps women become stronger. In the final analysis, only when women possess political and economic powers equal to men will an unambiguous message go out across the land: fantasize whatever you please, but nonconsensual sex will not be tolerated!

SELF-ASSESSMENT

I'm sure you've already learned a great deal about your fantasies, including what they mean and how you use them. Chances are that at least some of your fantasies focus on things you would like to try in the future or have enjoyed in the past. The obvious connection between these images and real life should concern you only if you're contemplating doing something nonconsensual or genuinely dangerous, such as engaging in unprotected intercourse with a nonmonogamous partner.

It is particularly important to make a distinction between thought and action when a fantasy violates your values or would be destructive or hurtful to yourself or others if acted upon. Contemplate these two questions:

- How comfortable are you with the knowledge that a dimension of your eroticism exists only in fantasy?
- How large a gap can you tolerate between what you imagine and what you choose to do?

No matter how scrupulously we differentiate fantasies from actions, one thing we know for sure: the erotic mind refuses to be pigeonholed. Therefore, most of us will encounter situations in which we allow unacceptable fantasy desires to affect our behavior. A middle-aged woman who fantasizes seducing pubescent boys may be tempted to try it if such a boy takes an interest in her. Similarly, a doctor who is turned on by the vulnerable beauty of one of his patients might artificially extend a naked examination and convince himself that she's enjoying it.

No one achieves anything like perfect erotic health. Some obviously stumble badly, to the detriment of themselves and others. However, those who conscientiously confront their mistakes have a chance to use them as opportunities for growth. Make a point of recalling situations in which compelling fantasies pulled you toward actions you later regretted. Use these memories to help define the conditions under which fantasies and action can interact positively as opposed to those in which a clear separation is essential.

NURTURING CHILDREN'S SEXUALITY

I once thought it odd that Eros is depicted as a small child. I now believe Cupid's youth symbolizes the fact that the seeds of adult eroticism are sown in childhood and adolescence. Not only is our capacity for joyous sensuality rooted in early experiences of positive touch, but our ability and willingness to give and receive affection is similarly linked with how we were held, caressed, and cared for as infants and small children. These are the foundations upon which our one-of-a-kind eroticism begins its extended development.

Although we are exposed to dangers and hurts throughout our lives, the most serious damage is often inflicted, with or without conscious intention, upon the young. I'm not simply referring to the devastating effects of overt abuse or neglect but also to what happens when a child is consistently prevented from following his or her natural curiosity or taught that pleasurable sensations are to be feared rather than enjoyed.

Like all living organisms, humans are equipped to survive even in harsh or barren environments. But those who will ultimately thrive require at least a small patch of emotionally fertile ground. Adults who are close to children—especially parents and other close relatives, teachers, the clergy, and counselors—are responsible for providing that fertile ground. *Erotically healthy people who are involved with children take an interest in their sexual development, especially the promotion and nurturance of positive, self-affirming attitudes toward sex.*

Whether we realize it or not, all of us who are close to kids are sex educators, a responsibility that involves so much more than disseminating facts. Of course, it's helpful when parents give simple, straightforward answers. Also important is well-planned sex education in our schools—beginning in the early elementary grades and expanding for adolescents and young adults. What matter most, however, are the everyday messages we give our kids about their worth, the value of their bodies, and the importance of their sexuality. These messages are communicated most powerfully through touch and direct observation. There is no better sex education, for instance, than observing an obviously affectionate bond between one's mother and father.

Other relatives and intimate family friends should also recognize they have subtle yet valuable contributions to make. Among the advantages of extended families is their ability to compensate for the fact that parents can't possibly be perfect. An admired older sibling, a grandparent who exudes unconditional love, or a trusted neighbor can often be easier to confide in than one's own parents.

The pivotal sex talk that parents are supposed to have with their kids is all but irrelevant. By the time it takes place, unspoken attitudes are already formed. And unless these discussions are part of an ongoing dialogue, they're inevitably awkward for everyone involved. Children absorb what we truly believe through our unrehearsed comments and behavior. Sex education at home is cumulative and has little to do with self-conscious discussions and speeches that mostly convey the message that sex is uncomfortable.

As crucial as it is that adults create a nurturing environment for children's sexual development, it's just as important that we avoid meddling in their sensual and sexual experimentation unless we have reason to believe they might be hurt emotionally or physically. Children have a right to sexual privacy. The process of building their eroticism belongs totally to them. Kids don't want to be asked about sex play with age-mates or themselves. If these activities are happened upon, however, we can reassure them they're normal and that they needn't fear punishment. I have the distinct impression that children who receive such simple reassurances frequently remember them gratefully as adults.

It is hazardous folly to believe we can protect our children from exploitation by teaching them to mistrust their sexuality. What they primarily need to know and believe is that *they* are the guardians of their bodies with the right to enjoy its pleasures and to speak up when touches don't feel good—whether that means Aunt Helen's sloppy kisses or a molester's furtive fondles.

We must teach children at least as much about their strengths as about their vulnerabilities. When we encourage them to act more innocent than they really are, we do them a grave injustice. Of course, we fervently wish that our kids could inherit a less risky world. But with deep-rooted self-worth, clear information, consistent care—and a little luck—they will be equipped to cope with the hard realities of sex and love.

SELF-ASSESSMENT

Unless you come to terms with the unfinished business of your own childhood, you will unwittingly pass along the same erroneous lessons about sex you learned long ago. Think about how your own sexual development could have been more effectively nourished. What do wish your parents or other sex educators had explained or done for you? What do you wish they *hadn't* said or done? If you could undo their mistakes, which ones would head your list?

It's one thing to have complaints but quite another to avoid repeating them. The best way to start is to treat yourself as you wish others had treated you. Then you will have a new foundation from which to nurture. It will help if you write down what you would most like to communicate about sex to the children you love. At the very least, lay out a few basic principles for how you intend to foster your child's sexual development. A variety of thoughtful books can guide you.[5]

APPRECIATING EROTIC DIVERSITY

No one who explores honestly the innermost realities of his or her eroticism finds complete "normalcy." Show me a person who claims to be free of sexual quirks and eccentricities and I'll show you someone tragically inhibited, more robotic than human. When we deny or reject our sexual idiosyncrasies we renounce who we are.

Erotically healthy people accept and appreciate their sexual uniqueness rather than fearing or fighting it. Our quirkiest quirks typically show up in the theater of the mind, so accepting them is easier for those who have successfully differentiated fantasy from behavior. To the extent that we feel safe and sexually free we will also reveal some, but probably not all, of our true preferences to our partners. Conversely, when we feel obligated to measure up to our partner's real or imagined expectations, we censor ourselves and thereby limit the amount of pleasure we can give and receive.

The ability to appreciate our own erotic individuality is interconnected with our willingness to accept variation in others. *Erotically healthy people appreciate sexual diversity in others as well as in themselves.* Each of us has a diversity comfort zone that determines how many and what kinds of sexual differences we can tolerate. Most of us find that our comfort zone enlarges as we get to know someone better. Research has shown this to be particularly true when it comes to sexual orientation. Those who hold even strongly disapproving attitudes about homosexuality usually begin to make subtle or even dramatic shifts in perspective when someone they care about comes out of the closet.

A similar process operates with other nontraditional manifestations of eroticism such as an interest in leather, casual sex, body piercing, tattoos, bondage, or cross-dressing, to name just a few. Generally speaking, the more acquainted you become with a particular erotic twist, especially if you know someone involved in it, the less charge it will carry. Familiarity fosters acceptance at least as often as it breeds contempt. These days titillating TV talk shows help give human faces to unusual sex practices. Unfortunately, too often these programs create a freak-show atmosphere rather than promoting genuine understanding.

No matter how large your diversity comfort zone becomes, there will always be sexual behaviors you simply cannot accept. If you've taken the time to clarify your values in the context of your background and personality, you probably have a good idea why you feel as you do. And I hope this book has shown how behaviors that make you uneasy can still teach you about the erotic mind—if you temporarily adopt a nonjudgmental attitude.

When certain sexual practices trigger visceral negative reactions, but you're not sure why, odds are you're responding to unquestioned assumptions and prejudices rather than consciously chosen values. It takes considerable courage to challenge nonrational beliefs, but it is also one of the most effective ways to stretch your diversity comfort zone. Why invest the effort? Don't accept the sexual quirks of others to make life easier for *them*—although this is surely a noble motivation. Embrace sexual diversity because it's your best hope for finding an abiding ease with your own erotic potential.

11

EROS FULFILLED

Profoundly subjective responses to peak turn-ons reveal the rewards of the erotic adventure.

When I developed the Sexual Excitement Survey, the dynamics of arousal were my primary concern. As I analyzed The Group's stories, however, I couldn't help noticing a striking similarity in virtually all of them, regardless of content: peak erotic experiences are as fulfilling as they are arousing. Whereas excitement and fulfillment can be quite different from each other, in peak sex they are virtually indistinguishable. This discovery opened up an unexpected opportunity to use the SES to explore the nature of sexual satisfaction.

When Dr. Maslow wrote about peak experiences he naturally emphasized the highly personal experiences of his subjects rather than the specific details of what took place. He knew that a vast array of events and situations can be catalysts for peak experiences. He also realized that it's not the events themselves but the individual's inner responses that produce the joy of a peak moment. Maslow's insight turns out to apply equally to peak eroticism. Unless we have a profoundly personal response, even the sexiest partner or situation will ultimately be little more than interesting—and probably not very memorable.

Once I realized that The Group's stories were valuable sources of information about fulfillment as well as excitation, I noted what respondents said about the subjective experience of peak turn-on. In addition to all the juicy details, The Group spontaneously mentions these personal responses far more frequently than any others:

Sensual and orgasmic intensity

Reduced inhibitions

Validation given and received

Mutuality and resonance

Transcendence of personal boundaries

Taken together, I believe these responses are the *essence* of peak eroticism. They represent the hopes and needs we bring to an erotic adventure and, just as surely, the keys to our fulfillment. As I describe each of them, notice which ones are familiar to you.

SENSUAL AND ORGASMIC INTENSITY

The sexual experience is, at the most fundamental level, an expression of your physical self. Without your inborn capacities to receive and process sensual stimulation, to build up muscle tension as you become aroused and to release it through orgasm, eroticism as you know it could not exist. During peak sex your body and all its senses spring to life.

Your response to touch is particularly likely to become more acute. If someone were to observe the stimulation you receive during peak sex they probably wouldn't notice anything out of the ordinary—a stroke here, a lick there. But from a purely subjective point of view, your receptivity goes into hyperdrive, bringing a richness to stimuli that under more normal circumstances might seem mundane.

Ironically, the flood of tactile sensations you experience during peak sex is *not* primarily the result of heightened physical sensitivity. Researchers who measure the physiology of arousal note that our skin sensitivity actually *drops* as excitement escalates. You feel so much more because you become totally absorbed in whatever is turning you on while screening out all extraneous stimuli. Although sexologists still have a lot to learn about it, this narrowing of focus is an altered state of consciousness quite similar to hypnosis. You might think of it as a "sexual trance."[1]

Forty-six percent of The Group's peak encounters and 39 percent of their favorite fantasies contain spontaneous references to the intensity of their sensations or their orgasms or both. And they don't just say, "That felt good." They rave about their sensations, making comments such as:

"My body came alive."

"Every touch made me tingle."

"I became lost in the texture of her skin."

"The sight of his naked body was too exciting to describe."

"I didn't think I was capable of feeling so much."

Women are more likely than men to make such comments (half of the women compared to 40 percent of the men). The three subgroups most likely to mention the intensity of their sensations and orgasms are lesbians (64 percent), bisexual women (63 percent), and bisexual men (60 percent).

People often mention being in tune with their "animal nature" during peak turn-ons. This connection with the instinctual, noncerebral world of the body not only brings with it a sense of strength and power, but also unrestrained feelings of joy. It's not unusual for men and women to be surprised by their own sexual capabilities and stamina. The liberal use of superlatives frequently makes storytellers sound like athletes describing those ineffable moments when they are so in touch with their bodies that extraordinary feats become effortless. But what stands out most is the total lack of struggle and the ease of it all.

For some, especially women, the ability to experience several orgasms within a brief period of time, or throughout an extended lovemaking session, becomes a symbol of their extraordinary responsiveness. Alexandra's encounter features multiple orgasms and an attentive, encouraging partner:

> I was having a relationship with an older man. One time we were making love and he was totally focused on me. As he stroked and manipulated my clit, he also kissed my breasts and sucked my nipples. He held me tight when I came which immediately made me come again. Then he finger-fucked me with such intensity that I came *again.* He said, "Come on darling," knowing full well that I had another orgasm stored in me. After I went down on him and he came in my mouth, he concentrated on my "encore" and I climaxed one last time. Whew!

Although multiple orgasms are fairly common during peak sex, they're by no means required. Nonetheless, when it comes to sensual and orgasmic intensity, if a little is good, more is usually better.

Sometimes, though, an encounter stands out not because of its intensity, but rather because it contains a gradual discovery of simple pleasures previously hidden or actively avoided. Harold, age fifty-seven, describes the quiet drama of such a moment:

> I don't like to admit it, but I've never liked being touched. Partly it's because I'm ticklish. I know this sounds strange, but sometimes my wife's touch reminds me of when my mom would try to hug me when I was a teenager and it felt awkward and I wanted to escape. As a man I prefer to have sex mostly with my genitals. Leave the rest of me alone. I'm just not the touchy-feely type.
>
> One morning my wife started caressing me as I was waking up. The night before we had one of our most honest conversations. I don't know if that was it, but as she touched me I knew it was okay to relax even though I still felt like cringing at first. Gradually (she must have been shocked as hell) she touched all the areas I normally hate.
>
> It was as if a layer of padding that had always covered my body fell away. Instead of recoiling from her or feeling ticklish, the entire surface of my skin was electrified and I *liked* it! I have no idea how long this lasted. I'm not even sure if I had an erection—for once it didn't matter. I know we didn't have intercourse. All day I wanted to enjoy my body. We worked in the yard, went for a walk and lounged in the sun, holding hands. My wife's amazement was nothing compared to mine. For once I *had* a body.

For Harold to drop the protective shield he had built up, he had to overcome a lifetime of uneasiness about the meanings and intentions associated with touch. I believe his newfound enjoyment of his body is a remnant of the sensuality of his infancy or early childhood—before fear crippled his capacity for pleasure. Though we may not be aware of it, the deep satisfaction of memorable sex often springs from a similar reconnection with an earlier, uncontaminated—innocent, if you prefer—delight in our physical selves.

REDUCED INHIBITIONS

If you're like most people, sex is best when you can throw yourself into it with abandon, unfettered by the vast array of restrictions, fears, worries, conflicts, and pressures that sometimes conspire to restrain you. We're very much a part of ancient cultural-religious traditions in which the body is associated with our primitive animalism. We're taught, not necessarily with words, that it must be tamed using the "higher" aims of the mind and spirit. Long before you reached adulthood, chances are you had unconsciously incorporated at least some of this anti-body bias into your personality. Even if you now embrace the *idea* that your desires for physical pleasure are natural and good, you may still *feel* conflicted about it.

Without fully realizing it you may also have learned to restrict the free flow of erotic energy through a process Wilhelm Reich called "muscular armoring."[2] If the postural and muscular patterns of your caretakers nonverbally communicated discomfort with physical pleasure, you probably imitated what you observed. To the extent that you still lock yourself into frozen postures, constrict your breathing, and curtail your range of motion, you narrow your capacity for sensual enjoyment and thus continue to obey the dictates of your training.

In addition to whatever limits you place on your body, you've also seen how emotional concerns, conflicts, and fears as varied as fingerprints are woven into your erotic patterns. Any deeply held doubts you may have about your desirability, attractiveness, sexual adequacy, or the normalcy of your preferences are particularly potent and persistent inhibitors. Your CET tries valiantly to transform these difficulties into aphrodisiacs but isn't always successful.

Miraculously, during peak experiences the body tends to release its rigid postures as all emotional concerns spontaneously evaporate. Somehow you're able to revel in the joy of the moment and celebrate your childlike freedom. Here's how Maslow describes these extraordinary states:

> . . . a complete, though momentary, loss of fear, anxiety, inhibition, defense and control, a giving up of renunciation, delay and restraint. The fear of disintegration and dissolution, the fear of being overwhelmed by the "instincts," the fear of death and insanity, the fear of giving in to unbridled pleasure and emotion, all tend to disappear or go into abeyance for the time being.[3]

One-third of The Group's peak encounters and a fifth of their fantasies spontaneously include references to themselves or their partners or both letting go of inhibitions, becoming playfully experimental, or highly expressive. I'm convinced that at least *some* reduction in those things that normally hold us back is a prerequisite for peak sex.

BETTER SEX THROUGH CHEMISTRY?

It's not unusual for people to seek chemical assistance in their struggle with inhibitions. Among The Group, 46 percent reported having at least one drink before peak encounters, 14 percent used marijuana, 4 percent used cocaine or other stimulants, and a handful used psychedelics or nitrite inhalants ("poppers"). Any of these drugs can help overcome inhibitions and increase our focus on building sensations. The risk, of course, is that we may become dependent on one or more drugs to free us. In the worst cases it's difficult or impossible for a person to experience high arousal when sober—obviously a sign of serious trouble.[4]

Relatively few members of The Group who use alcohol or other drugs before or during their peak turn-ons actually mention them in their stories. Instead they reveal their drug use by answering questions about it in the SES. Those who do allude to drugs in their stories usually mention drinking as part of the social context for an encounter or as a calculated strategy for overcoming their own inhibitions or those of a reticent partner. The smaller number who mention smoking marijuana in their stories typically comment on its ability to intensify sensations.

The relatively few instances in which drugs play a central role in a peak turn-on invariably involve one of the hallucinogenic substances such as LSD ("acid"), psychoactive mushrooms, MDMA ("ecstasy"), or hashish (a highly concentrated form of marijuana). These powerful drugs so radically alter the perceptions and emotions of whoever is under their influence, they're often seen as coparticipants in the experience. Note how the drug ecstasy is pivotal in Jennifer's story of love, sensuality, and obliterated inhibitions:

> My lover and I were vacationing on a beach in Mexico. I wanted him to be in love with me as I was with him, but he was holding back. One afternoon we took ecstasy together. The drug had an almost magical effect, helping us to open our hearts and express feelings that were locked inside. It was overwhelming to learn that he was madly in love with me but also terribly afraid of being hurt again (his ex-wife had dumped him for a younger man).
>
> Every touch was a revelation, as if we were discovering each other for the first time. Our senses were drinking in the wind, the sand between our toes, the rhythm of the surf and the penetrating warmth of the sun. We were even more overcome by emotions as we kissed and talked for hours.
>
> Back at the hotel my excitement continued to be mostly emotional. He let himself become vulnerable to me. I never realized this man felt so much inside. His openness was a gift that made me want to make love to him without reservation. It was beyond comprehension. I was amazed by how we were both freed of all inhibitions. Verbally and physically he expressed adoration of my body and soul.
>
> He took me strongly in a virile, manly way and I also ravaged him. We naturally enjoyed acts we would have avoided before. We knew we could be nasty and still be

> loved. I knew he cherished me as much as I did him. More than once I cried at the beauty of the moment—the curve of his earlobe, the softness of my skin against his lips. We were all heart, all soul, and all body.

Clearly the drug was a catalyst for a remarkable degree of sensual and emotional freedom and intimacy. Naturally, you might wonder if all this ecstatic love was merely chemically induced. Jennifer had a similar concern:

> When I awoke the next morning I immediately wondered if Eric would be cautious again. Did he *mean* what he said on ecstasy? Even though we were completely down from the drug, Eric kissed me, held me, and softly told me how much he loved me (he hardly ever said these things before). We live together now and have discussed this experience often. We both agree that the drug removed our inhibitions so we could be completely truthful. The intensity was much greater than normal but the feelings were real—because we still feel them.

Had Jennifer and Eric needed to take ecstasy on a regular basis to sustain their intimacy, the quality of their relationship would eventually have deteriorated until there was nothing left but an artificial high. Luckily, their drug experience was simply a tool that worked for them.

SURPRISE AND SEXUAL FREEDOM

Of course, inhibitions release their grip in many other circumstances. As you may recall from Chapter 1, surprises are a common memorability factor. One marvelous aspect of a sexual surprise is that it obliterates our inhibitions and catches us with our defenses down. Sean, a gay man in his early twenties, demonstrates this process with an ironic twist:

> I had an incredible experience *after* I broke up with my lover of three years. Ours had been the first relationship for us both so it was quite tumultuous. Even though we were no longer an "item" we still saw each other around town. One

time we stopped to chat, went out for coffee, and, to my total amazement, eventually ended up in bed—the first time in over a year. Our sex life had never been easy, but this time I was so comfortable with Bill I couldn't believe it. I didn't worry about what he thought of me, or how I looked, or whether he liked what I was doing, or when we were going to come, or anything! None of that mattered at all.

After sex we lay in bed watching the sunset. At one point I was almost shocked to realize his hand had been resting on my dick and mine on his for a long time. I don't remember ever being so exposed yet so relaxed. We commented how funny it was that we would finally be comfortable with each other now that our relationship was over. It was sad but also very positive. I not only let go of my sexual inhibitions that day, I also gave up most of my resentments. It certainly made our breakup a lot less painful. We're still close friends.

Sean's unexpected freedom with Bill illustrates the paradoxical nature of inhibitions. Even though it's easy to assume that we'd all be much better off without them, sexual inhibitions often play a positive role in memorable sex by providing the reticence and restraint out of which freedom can spring. When inhibitions loosen their grip they release a burst of energy and vitality. Again and again The Group's stories illustrate this dynamic interaction of inhibition and excitation.

VALIDATION GIVEN AND RECEIVED

Whenever I discuss the rewards of peak eroticism with my clients, sooner or later it usually comes down to this: satisfying sex leaves the participants feeling affirmed, highly desirable, and worthwhile—deeply validated. Focus your attention on the warm glow of fulfillment following a peak turn-on and you'll probably notice a good feeling about yourself and an unmistakable appreciation of your partner. If you and your partner also care for each other, the affirmation will mean even more.

In psychotherapy it often takes some gentle coaxing before clients shift their attention away from the details of a sexual experience and focus instead on feelings of increased self-worth or admiration for their partners. So I wasn't sure what, if anything, The Group might say about giving and receiving validation in their stories. As it turns out, without any prompting they regularly pepper their tales with spontaneous praise for their partners and themselves. Twenty-nine percent of their peak encounters contain clear statements of approval such as:

> "She was the sexiest women I've ever been with."
>
> "A masterful lover!"
>
> "I was really hot that night—and not afraid to show it."
>
> "He made me feel beautiful."
>
> "She was a precious jewel to me."

Women are more likely to make affirmative statements (35 percent) than men (23 percent). Lesbians and gay men make these sorts of comments about themselves or their partners with exactly the same frequency as they mention sensual or orgasmic intensity (64 percent of the lesbians and 35 percent of the gay men), indicating that for them validation is an exceedingly important dimension of fulfilling sex.

The Group is even more likely to give and receive affirmation in their favorite fantasies (42 percent), with men more likely to do so (46 percent) than the women (37 percent). In fact, giving or receiving validation is the single most frequent subjective response to fantasy among The Group as a whole and all the male subgroups, including straight men (43 percent) and gay men (67 percent). I don't know why, but lesbians include affirmative statements in their fantasies only about half as often as they do in their stories of actual encounters.

Remember, I didn't specifically ask The Group how they felt about themselves or their partners. These are completely spontaneous expressions. I'm convinced that truly satisfying encounters and fantasies virtually always bolster the worth of the participants. The high degree of self-disclosure in positive sex—which reveals your nakedness in more ways than one—increases the significance of whatever responses you give or receive.

Connie, age twenty-nine, offers a powerful example of the role of affirmation in a loving encounter that also marks the transition to a deeper level of caring:

> After attending a lecture, Marv returned to my apartment for tea and talk. The feeling was cozy and intimate. I asked him to sleep over but said I didn't want to have sex. At first he said he'd leave because it would be too frustrating to sleep with me and not have me but then he changed his mind and stayed.
>
> Up to this point our relationship had been mostly sexual with a little personal conversation. The lecture, discussion, and willingness to sleep together nonsexually was a new dynamic. In retrospect, I think part of me was testing to see if he wanted me for more than sex, though I wasn't aware of this at the time.
>
> Once in bed we kissed good night and a warm feel flowed through me like heated olive oil. I felt a new level of respect, appreciation, and being cared for. He immediately sensed my openness and his sensitivity turned me on all the more. We caressed each other slow and easy. Of course, I ended up taking him inside me and having a furious orgasm. He was also very moved by the experience. Lying there quietly with him in the wet heat afterward was sweet, so sweet.

Just as we might expect, Connie's sense of being accepted and affirmed by Marv is connected to the developing intimacy between them. However, analysis of The Group's peak turn-ons reveals that even casual encounters or fantasies can be profoundly validating. Raoul, an interior designer in his early thirties, describes the affirmative impact of a one-time encounter:

> Some of my friends tease me because I don't like one-night stands—in fact, I hate them! Maybe that's one reason why I have trouble meeting women (besides the fact that I'm not really over my last relationship). One evening after working late on a project that wasn't going well, I went out for a drink. I was in one of those moods when I question my abilities and generally rip myself apart. I finished a drink and was about to leave when the bartender brought me another,

pointing out an attractive woman across the bar who had bought it for me. I went over to talk with her even though I was eager to get home.

Not only was she beautiful with a radiant smile, she was also smart and funny—and married. She was in town for a conference and this was her last night. We really hit it off so she invited me to her hotel. I was naive enough to think we would just continue talking. She made me feel so good about myself, telling me I was handsome and that I must be a great designer because I dress with such style. In a way these sounded like empty lines, but she seemed sincere and I needed to hear it.

I spent the night with her and have never felt so good about casual sex. Even though we hardly knew each other it was very intimate. I can't remember ever being a better lover. In the morning when I took her to the airport she softly said, "I'll always remember you, my darling," with such finality and love that I cried on the way to my office. It wasn't so much that I wouldn't see her again but that she somehow made me feel worth something. By midafternoon I had worked out my problem at work. Maybe this was pure coincidence but I don't think so.

To comprehend the importance of validation in peak eroticism, keep in mind the role of self-doubt in the sexual scenarios that most excite us. One function of our CETs is to help us demonstrate our worth and desirability and counteract lingering negative beliefs about ourselves. Part of the reason peak sex is so deeply satisfying is that it gives a potent boost to our self-esteem, as it clearly did for Raoul. You've no doubt seen popular caricatures in which a post-orgasmic lover inquires, "Was it as good for you as it was for me?" We may be too sophisticated to make such a blatant request for approval, but that doesn't mean we don't want it.

MUTUALITY AND RESONANCE

Modern books about sexual problems and enhancement warn of the dangers of assuming that your partner knows how, where, and when you like to be touched. Sex therapists, usually an unflappable bunch, cringe when they hear their clients express two commonly held beliefs: "If you loved me you would know what to do" and "If I have to ask for it I don't want it." Such beliefs are setups for disappointment and frustration. Nevertheless, chances are you've had such thoughts yourself despite knowing, at least intellectually, that they're unrealistic.

An important reason for the persistence of such thinking is the fact that you've probably had encounters—and certainly fantasies—in which your partner knows, as if by magic, exactly what pleases you. More often than not, peak encounters have at least some of this quality of perfectly meshing timing, touch, and rhythm. About a quarter of The Group's encounters and a fifth of their favorite fantasies specifically mention this sense of being highly in tune with their partners—I call it mutuality and resonance. These are the sorts of comments they make:

"We were on the same wavelength."

"Everything I wanted she wanted too."

"He read my body like a book."

"It was as if my every secret desire was obvious."

"Our movements were perfectly synchronized."

"I played him like a violin and he loved it."

Women are significantly more likely than men to mention mutuality when describing their peak encounters (33 percent of women compared to 19 percent of men). Women are also more inclined to make mutuality a component of their fantasies (24 percent of women versus 17 percent of men). Lesbians are the most likely of all to refer to resonating with their partners, mentioning it in 45 percent of their encounters and in half of their fantasies.

Women are taught, directly or indirectly, that the right man will know more about her sexuality than she does. Outside the sexual arena, girls and young women learn to place a high value on being sensitive and responding to the needs of others. Both factors contribute to women's propensity to look for reciprocation with a perfectly matched partner.

Alice, now age forty-nine, recalls an experience from eight years earlier in which synchronized movements produced an exquisite nonverbal resonance:

> This was the second time my lover and I had sex. I was living and working in Paris. He was an Iraqi studying for an advanced degree. We met at a dance and I was impressed by his natural grace and rhythm—a superb athlete. On this occasion we danced to Greek folk music at home. As our movements became more sensuous our bodies swayed in fluid unity. We continued dancing, gradually removing each other's clothes until we were stark naked. His penis was between my legs and I thrilled to its hardness and warmth.
>
> When the music ended we fell to the floor. He came into me slowly and with the same perfect rhythm. I arched up so he could lick my nipples. We continued to move in harmony until we both climaxed. We could hardly be more different in terms of culture and background, but we were effortlessly connected with each other. What he wanted was what I wanted was what he wanted.

Among The Group's tales of peak sex there are a number in which music and dance are avenues to heightened mutuality. Others mention extended holding, kissing, and caressing—often while breathing in unison—as practical methods for establishing resonance.

Although men don't mention it as frequently as women, they unquestionably appreciate the same feeling. Both men and women yearn for opportunities to bridge the chasm that separates all of us into distinct, ultimately lonely, individuals. In peak eroticism lovers find a common playground in which to express their complementary desires. If only for a moment, the fundamental loneliness of being human is relieved.[5]

TRANSCENDENCE OF PERSONAL BOUNDARIES

In peak sex you may become so self-expressive, so clear about who you are and what you want, that your sense of self actually expands. You may move beyond the confines of habit and identity and enter an altered state of consciousness known as transcendence.

In their stories of peak sex, over 10 percent of The Group uses the poetic language suggestive of transcendence:

> "I felt a part of all that is."
>
> "The whole universe became erotic."
>
> "It was the primal dance of life."
>
> "Our love was unique yet infinite."

Because it's so difficult to describe a transcendent experience there are undoubtedly others who feel it but don't know how to translate their experiences into words. Transcendence connects us with the great mysteries of life. Our natural response to genuine mystery is usually to remain silent. When we try to explain what has touched us we run the risk of distancing ourselves from the experience because our descriptions are inadequate to convey its fullness.

Analysis of stories with transcendent aspects reveals several recurring themes:

- A feeling of participation in the grand scheme of existence.
- A clear though often inexpressible sense of meaning and purpose.
- A sense of completion, not needing anything else, and an acceptance of what is.
- An integration of opposites, most notably a healing of the traditional mind/body split and a bridging of the gap between masculine and feminine.[6]

Those who have transcendent experiences during peak sex usually don't consciously seek them out. Instead these experiences unfold naturally from a level of participation so all-encompassing and free that the self is unexpectedly released from its normal constraints. Carla's story captures the feeling:

> My boyfriend and I had been driving around all day, exploring new and exciting places. It began to rain and soon thun-

> der and lightning followed. We both had the same idea: let's get a bottle of champagne and drive up to Twin Peaks to watch the city being drenched by this incredible storm.
>
> We cheered when we found no one else there. As we sipped our champagne and talked there was an uncanny chemistry between us. Sometimes we would talk, sometimes kiss, sometimes be silent, holding hands, listening to the rain as it beat upon the windshield, watching and feeling it seep through the slightly opened windows. Our breath was taken away when bolts of lightning lit up the sky and thunder rumbled.
>
> Soon we were embracing. Our kisses matched the rhythms of the storm. When we made love it seemed so much more beautiful than any time before. Maybe it was the awesome power of the storm combined with the fact that we could be caught at any moment? All I can say is that we both felt a high afterward (and it wasn't the champagne because we had very little). The feeling was more mystical, meditative. I walked a bit lighter for several days.

Theirs is a marvelous synchronicity of time and place, a union of love and lust joined with the majestic forces of nature. During transcendent experiences, with or without an overtly erotic component, the perception of oneself as a part of the natural world is often the catalyst that draws the participants into the larger universe.

Surely the spirit of adventure that Carla and her boyfriend shared throughout the day also primed them for the ecstasies to follow. This is definitely not a safe encounter. Whereas most of us would retreat indoors at the first sign of a thunderstorm, these two seek out a mountaintop, eager to be closer to and revel in the danger and drama that moves them so deeply. In peak eroticism, as in all of life, lucky moments often happen to those willing to go out on a limb.

Transcendent moments sometimes lead to growth because they shake up and enlarge our perceptions so that apparent contradictions can be tolerated and integrated. Like Carl Jung before him, Mihaly Csikszentmihalyi has pointed out that in growing we simultaneously become both more separate *and* more interconnected:

> Following a flow [peak] experience, the organization of the self is more *complex* than it had been before. It is by becoming increasingly complex that the self might be said to grow. Complexity is the result of two broad psychological processes: *differentiation* and *integration.* Differentiation implies a movement toward uniqueness, toward separating oneself from others. Integration refers to its opposite: a union with other people, ideas and entities beyond the self. A complex self is one that succeeds in combining these opposite tendencies.[7]

When we surrender to a transcendent experience, we glimpse our universal aspects, moving beyond the limitations of the ego and its illusions of separateness. Yet the great paradox of transcendence is that while self-consciousness totally disappears, we know more clearly than at any other time exactly who we are.

TRANSCENDENCE AND SPIRITUALITY

Transcendence and spirituality are closely related. Those who have some familiarity with spiritual practices—meditation, for instance—or who have had mystical or religious experiences may be more likely to recognize moments of transcendence in their erotic lives.

Unfortunately, for our ability to recognize the spiritual aspects of eroticism, in most of the world's great religions, transcendence is of the spirit, whereas the joys of the body occupy a lower realm. Therefore, if you have been influenced by the teachings of virtually any organized religion, you probably believe, subconsciously at least, that sexual ecstasy and spiritual awareness are incompatible.

Even though transcendent eroticism usually happens quite by accident—an unexpected gift—certain techniques can help anyone actively explore the mystical dimensions of the erotic.[8] One of the best-known approaches is Tantra. Both a vision and set of practices, Tantra is said to have developed thousands of years before Christianity as an offshoot of Hinduism in India. Initially, it was a reaction to the belief that the denial of sexuality was necessary for enlightenment (sound familiar?). Tantra recognizes eros as a vital life force. It seeks to mobilize and shape—rather than to suppress—erotic energies as a pathway to the divine.

Although many Westerners find their ideas too strange to be of interest, Tantric practitioners believe that ecstasy is most likely to occur when relaxation and high states of excitement are combined. Rather than having tension build until it culminates in orgasmic release, Tantra calls for *relaxing into arousal.* It advocates recirculating erotic energy for extended periods, which sometimes results in orgasms that are long-lasting and not particularly genitally focused.

If we weren't so conflicted about eroticism I have no doubt that many of us would be much more cognizant of its transcendent and spiritual aspects. Peak erotic experiences are perfectly suited to transcendence because they engage us totally, enlarge our sense of self by connecting us with another or with normally hidden dimensions of ourselves or both, and expand our perceptions and consciousness. What a pity so few of us are encouraged to discover all of eros's gifts.

YOUR PERSONAL RESPONSE STYLE

No two people respond exactly the same way even to very similar peak experiences. Your background and individual propensities cause you to focus on certain responses while ignoring or downplaying others. This is the order in which The Group mentions their subjective responses, starting with the most frequent:

Sensual and orgasmic intensity

Reduced inhibitions

Validation given and received

Mutuality and resonance

Transcendence of personal boundaries

You may have felt some or all of these responses, or different ones not on this list. To learn more about what erotic fulfillment consists of for you, make your own list of responses, starting with the ones you recall most frequently and then adding those you recall experiencing less often. The responses toward the top of your list obviously play a crucial role in helping you feel fulfilled.

The next time you're engaged in an enjoyable encounter or fantasy, make a point of observing and savoring those subjective responses that stand out. You may make these observations during the experience or afterward. Either way, there's no reason for this process to make you self-conscious because you're not judging or evaluating—just observing. When you focus on your favorite responses you can experience more fully the things you already like.

However, if you make a point of noticing the responses with which you are less familiar, you can open up the possibility of a qualitative shift in your experience. It's your *unfamiliar* responses that may offer the greatest opportunities for developing your erotic potential.

As an experiment, pick one or two responses at the bottom of your list. Move into a comfortable position, close your eyes, and recall one of your favorite encounters or fantasies. Imagine how your experience might have differed had you been more aware of these responses. Is it possible you could have enjoyed yourself even more had you been attuned to them? By repeating this simple exercise with a variety of different memories you can gradually train yourself to notice a wider range of responses, and in the process bring a new richness to your eroticism. But don't fall into the trap of trying to experience all responses equally; that would be unrealistic and pointless. After all, what could be more uniquely individualistic than the shape and texture of eros fulfilled?

THE PASSION-FULFILLMENT PARADOX

I hope this book has helped you make valuable discoveries about how the erotic mind creates and intensifies desire and arousal. I also hope you have cultivated the paradoxical perspective—a different way to explore your eroticism. It may not be the easiest way, yet I trust your willingness to tolerate contradictions and ambiguities has proven fruitful.

As you've explored the profoundest subjective rewards of the erotic adventure you've probably come face to face with one of the most fundamental of all erotic paradoxes: *even though passion and fulfillment have a close, reciprocal relationship, there is an unavoidable tension between them.* While the idea of perpetual fulfillment holds an undeniable appeal, the truth is that never-ending fulfillment, if there ever were such a thing, would ultimately lead to boredom—the polar opposite of passion.

Of course, passion seeks fulfillment as its greatest reward. In so many ways fulfillment enhances passion because it teaches us an enormous amount about the secrets of arousal. It is equally true, however, that fulfillment inevitably subdues passion because it quenches need, and thus desire. And without desire there is no reason for passion.

You might think of the passion-fulfillment paradox as part of the larger human drama in which satisfactions of all kinds sow the seeds of discontent. But for those who accept the ways of the erotic mind, passion and fulfillment are accurately seen as two essential parts of a whole. When you know what you want and are lucky enough to find it, you feel not only uplifted and enlivened, but also satiated—not a bad feeling at all. Yet it's just a matter of time until new desires begin to stir. You're dancing to the age-old rhythms of eros.

APPENDIX

THE SEXUAL EXCITEMENT SURVEY

Dear Reader:

I've included this modified version of the SES so you can see how I gathered information about peak sex for my research. More important, I hope you'll actively use it as a tool for expanding your self-awareness.

Part I is concerned with your most memorable real-life peak encounters, while Part II focuses on fantasy. Responding to all the questions can stimulate your thoughts. However, particularly crucial items appear in boxes. These open-ended questions ask you to write about your personal experiences in as much detail as you wish.

If you'd like to help expand my research, consider sending me your responses (anonymously, of course). This choice is completely optional, a decision you can put off until later if you prefer. *Only* if you send in your answers is it necessary to respond to Part III (personal background information).

If you decide to become a research participant, simply print or type the number of each question on sheets of paper. There's no need to write out the questions. Remember *not* to put your name anywhere. For questions with lettered or numbered choices, select the appropriate letter or number of your answer. For questions in boxes, please type or print your stories, using as many pages as you wish. Mailing instructions are at the end of the survey.

PART I: REAL-LIFE ENCOUNTERS

Think back over all your sexual encounters with other people. Allow your mind to focus on two specific encounters that were among the most arousing of your entire life. Describe each of them in as much detail as you wish.

1. Describe exciting encounter.

2. How old were you when you had this encounter?

3. What kind of relationship did you have with the partner(s) in this encounter?

a. Anonymous
b. Acquaintance
c. Boyfriend/girlfriend
d. Primary relationship or spouse
e. Multiple partners
f. Other (specify) ________________

4. What do you think made this encounter so exciting?

5. How would you rate your level of *excitement* during this encounter, especially compared to your usual ones?

Not exciting [-0-1-2-3-4-] Extremely exciting

6. How would you rate your level of *fulfillment* during this encounter, especially compared to your usual ones?

Not fulfilling [-0-1-2-3-4-] Extremely fulfilling

7. How important was each of the following six groups of emotions in this encounter? Within each group of feelings, base your rating on whichever feeling was *most* important. (Note: some emotions, especially the "negative" ones, may be very important even though they're not particularly intense.)

a. **Exuberance** (related emotions: joy, celebration, surprise, freedom, euphoria, and pride)

Not at all important [-0-1-2-3-4-] Very important

b. **Satisfaction** (related emotions: contentment, happiness, relaxation, and security)

Not at all important [-0-1-2-3-4-] Very important

c. **Closeness** (related emotions: love, tenderness, affection, connection, unity [oneness], and appreciation)

Not at all important [-0-1-2-3-4-] Very important

d. **Anxiety** (related emotions: fear, vulnerability, weakness, worry, and nervousness)

Not at all important [-0-1-2-3-4-] Very important

e. **Guilt** (related emotions: remorse, naughtiness, dirtiness, and shame)

Not at all important [-0-1-2-3-4-] Very important

f. **Anger** (related emotions: hostility, contempt, hatred, resentment, and revenge)

Not at all important [-0-1-2-3-4-] Very important

8. Before or during this encounter, which of the following drugs did you use? (note as many as apply)

a. None
b. Alcohol
c. Barbiturates/tranquilizers ("downers")
d. Stimulants (cocaine, "speed")
e. Marijuana
f. Nitrite inhalants ("poppers")
g. Psychedelics (LSD, "ecstasy," etc.)

9. Describe exciting encounter #2.

10. How old were you when you had this encounter?

11. What kind of relationship did you have with the partner(s) in this encounter?

a. Anonymous
b. Acquaintance
c. Boyfriend/girlfriend
d. Primary relationship or spouse
e. Multiple partners
f. Other (specify) ________________

12. What do you think made this encounter so exciting?

13. How would you rate your level of *excitement* during this encounter, especially compared to your usual ones?

Not exciting [-0-1-2-3-4-] Extremely exciting

14. How would you rate your level of ***fulfillment*** during this encounter, especially compared to your usual ones?

Not fulfilling [-0-1-2-3-4-] Extremely fulfilling

15. How important was each of the following six groups of emotions in this encounter? Within each group of feelings, base your rating on whichever feeling was *most* important. (Note: some emotions, especially the "negative" ones, may be very important even though they're not particularly intense.)

a. **Exuberance** (related emotions: joy, celebration, surprise, freedom, euphoria, and pride)

Not at all important [-0-1-2-3-4-] Very important

b. **Satisfaction** (related emotions: contentment, happiness, relaxation, and security)

Not at all important [-0-1-2-3-4-] Very important

c. **Closeness** (related emotions: love, tenderness, affection, connection, unity [oneness], and appreciation)

Not at all important [-0-1-2-3-4-] Very important

d. **Anxiety** (related emotions: fear, vulnerability, weakness, worry, and nervousness)

Not at all important [-0-1-2-3-4-] Very important

e. **Guilt** (related emotions: remorse, naughtiness, dirtiness, and shame)

Not at all important [-0-1-2-3-4-] Very important

f. **Anger** (related emotions: hostility, contempt, hatred, resentment, and revenge)

Not at all important [-0-1-2-3-4-] Very important

16. Before or during this experience, which of the following drugs did you use? (note as many as apply)

a. None
b. Alcohol
c. Barbiturates/tranquilizers ("downers")
d. Stimulants (cocaine, "speed")
e. Marijuana
f. Nitrite inhalants ("poppers")
g. Psychedelics (LSD, "ecstasy," etc.)

PART II: SEXUAL FANTASIES

The focus of Part II is your personal experiences with sexual fantasy, in the past as well as the present. A sexual fantasy is simply a mental image, daydream, thought, or feeling that turns you on. Fantasies can be brief and simple or long and complex. If you're unclear about what fantasies are, read the fantasy section in Chapter 1.

17. At what age do you first remember having a sexual fantasy?

18. Describe one of the first sexual fantasies you can remember.

19. Considering *all* your sexual fantasies that include other people, what proportion of the important characters—besides yourself—are of the same or opposite sex as you?

All same sex [-0-1-2-3-4-] All opposite sex

Following are a variety of statements about sexual fantasy. How frequently does each statement apply to you personally? For each statement, select a number from this scale that best reflects your experience:

Never [-0-1-2-3-4-] Very frequently

20. I fantasize about my past sexual experiences.

0 1 2 3 4

21. I fantasize about desired future experiences.

0 1 2 3 4

22. I fantasize about things that couldn't really happen.

0 1 2 3 4

23. I fantasize about things I wouldn't actually want to do.

0 1 2 3 4

24. I fantasize about someone besides my regular sex partner(s).

0 1 2 3 4

25. I fantasize when I masturbate.

0 1 2 3 4

26. I fantasize when I'm having sex with a partner.

0 1 2 3 4

27. I fantasize about sex with two or more partners at the same time.

0 1 2 3 4

28. I have fantasies when I don't want to.

0 1 2 3 4

29. I'm embarrassed or uncomfortable about my fantasies.

0 1 2 3 4

30. I think my fantasies are less interesting than other people's.

0 1 2 3 4

31. I wonder if my fantasies are normal.

0 1 2 3 4

32. I wish my fantasies were different than they are.

0 1 2 3 4

33. I've made a conscious effort to change my fantasies.

0 1 2 3 4

34. Imagine yourself really wanting to be sexually aroused but, for some reason, you're not. Based on everything you know about your sexuality, describe the fantasy that would be *the very most likely* to arouse you.

35. What are your ideas about what makes this fantasy so exciting? Please be as specific as you possibly can.

36. Describe the "climax"—the most intense point of excitement—of this fantasy.

37. How important is each of the following six groups of emotions in this fantasy? Within each group of feelings, base your rating on whichever feeling is *most* important. (Note: some emotions, especially the "negative" ones, may be very important even though they're not particularly intense.)

a. **Exuberance** (related emotions: joy, celebration, surprise, freedom, euphoria, and pride)

Not at all important [-0-1-2-3-4-] Very important

b. **Satisfaction** (related emotions: contentment, happiness, relaxation, and security)

Not at all important [-0-1-2-3-4-] Very important

c. **Closeness** (related emotions: love, tenderness, affection, connection, unity [oneness], and appreciation)

Not at all important[-0-1-2-3-4-] Very important

d. **Anxiety** (related emotions: fear, vulnerability, weakness, worry, and nervousness)

Not at all important [-0-1-2-3-4-] Very important

e. **Guilt** (related emotions: remorse, naughtiness, dirtiness, and shame)

Not at all important [-0-1-2-3-4-] Very important

f. **Anger** (related emotions: hostility, contempt, hatred, resentment, and revenge)

Not at all important [-0-1-2-3-4-] Very important

38. Think about *all* the different fantasies that excite you. What *percentage* of all your fantasies have a similar theme to the one you just described?

39. For *how many years* have you been aroused by fantasies similar to the one you just described?

40. How often do you use erotic materials—such as sexually explicit books, magazines, videos, etc.—either alone or with a sex partner?

Never [-0-1-2-3-4-] Very frequently

41. If you ever use erotic materials, what is the most common effect they have on you?

No effect [-0-1-2-3-4-] Highly arousing

42. Which of the following people have you told about your most exciting fantasy? (note as many as apply)

a. No one
b. A parent
c. A sibling
d. A friend
e. An acquaintance
f. A stranger
g. A casual sex partner
h. A regular sex partner
i. A therapist

PART III: PERSONAL BACKGROUND INFORMATION

Respond to this section only if you are mailing in your answers.

43. Your gender?

44. Your age?

45. Your occupation?

46. Your race?

a. Asian/Pacific Islander
b. Black
c. Hispanic
d. Caucasian
e. Other ________________

47. In which state do you live?

48. How would you describe the community in which you live?

a. Large city
b. Suburban
c. Small city
d. Rural

49. Your highest level of formal education?

a. Less than high school
b. High school graduate
c. Some college
d. College graduate
e. Some graduate work
f. Graduate degree

50. In which organized religion did you participate *as a child*?

a. None
b. Protestant
c. Catholic
d. Jewish
e. Other ________________

51. In which organized religion do you participate *now*?

a. None
b. Protestant

c. Catholic
d. Jewish
e. Other ________________

52. How much influence do you think your religious beliefs (past or present) have on your *current* attitudes and feelings about sex?

No influence [-0-1-2-3-4-] Strong influence

53. How old were you when you first masturbated?

54. How many times do you masturbate now in an average month?

55. If you masturbate, how many minutes do you usually spend during one session?

56. How old were you when you first had a feeling of sexual attraction toward another person?

57. How old were you when you first did *any* kind of sexual touching with another person?

58. How old were you when you first had an orgasm with another person (from any kind of stimulation)?

59. How many different sexual partners have you had *in your lifetime?* (any sexual contact, not necessarily intercourse)

60. During the last year, how many times have you had sex with a partner in an average month (any sexual contact, not necessarily intercourse)?

61. How many times would you *like* to have sex with a partner in an average month?

62. During the last year, how many orgasms have you had in an average month (by yourself and with a partner)?

63. When you have sex with a partner, about what percentage of the time do you have an orgasm?

64. What is your current marital/relationship status?

a. Single/never married
b. Married
c. Separated/divorced
d. In primary relationship, but not married

The next four questions are about your current primary relationship. If you are not involved in a relationship, please skip to question 69.

65. How long have you been involved in your current relationship?

66. Is your partner male or female?

67. How many times have you had sex with this partner in the last month (*any* sexual contact, not necessarily intercourse)?

68. Since you became involved with this person, with how many other partners have you also had sex (*any* sexual contact, not necessarily intercourse)?

69. How do you define your sexual orientation?

Exclusively homosexual [-0-1-2-3-4-5-6-] Exclusively heterosexual

70. Overall, how satisfied do you feel with your current sex life?

Not at all satisfied [-0-1-2-3-4-] Very satisfied

71. How would you rate your overall level of self-esteem?

Very low [-0-1-2-3-4-] Very high

Please answer yes or no for each of the following questions:

72. Before puberty, did you ever have any sexual contact (not necessarily intercourse) with an adult?

73. Have you ever had any sexual contact with a sibling?

74. Have you ever had any sexual contact with a parent or step-parent?

75. Have you ever been forced to have sex when you didn't want to?

76. Have you ever forced another person to have sex with you when he or she didn't want to?

77. Have you ever done anything sexually that was against the law?

If yes, what did you do?

78. Have you ever been arrested because of your sexual behavior?

If yes, what were you arrested for?

79. What was the total amount of time you spent filling out this survey?

80. Are there any comments you would like to make about this survey? (what you liked and didn't like about it, how it could be improved, etc.)

81. In the course of reading *The Erotic Mind,* when did you respond to this survey?

a. When I first read about it in chapter 1.
b. While I was reading the book.
c. After I finished reading the book.

If you've decided to send in your answers, please make sure you've answered every relevant question (partial responses cannot be included in the research) and that your name does *not* appear anywhere. Mail your responses to:

Jack Morin, Ph.D.
P.O. Box 1045
Belmont, CA 94002-1045

NOTES

Notes preceded by the 📖 symbol contain recommended readings.

INTRODUCTION: SEX AND SELF-DISCOVERY

1. Zilbergeld (1992), p. 320.
2. Schnarch (1991), p. 314.
3. See Masters and Johnson (1970).

1 PEAK EROTIC EXPERIENCES

1. Those interested in peak experiences should be aware of the work of psychologist Mihaly Csikszentmihalyi (1990). He uses the term "optimal experience" and calls that state "flow," which also is the name of his best book on the subject.

2. Maslow (1968), p. 103.

3. Most sex research based on surveys is quantitative, with the emphasis on performing statistical analyses. To yield significant findings, quantitative survey research requires relatively large numbers of respondents. The SES is a qualitative study that emphasizes thematic analysis of written stories rather than statistics. This type of research is well suited to smaller sample sizes.

4. Fifty-two percent of The Group are women and 48 percent are men. Eighty-three percent of the women are primarily or exclusively straight, 10 percent are lesbian, and 7 percent are bisexual. Three-quarters of the men are primarily or exclusively straight, 16 percent are gay, and 9 percent are bisexual.

Most are white (79 percent), but other ethnic groups are also represented, including Asians (9 percent), Hispanics (8 percent), and blacks (3 percent).

The Group averages about ten encounters per month, ranging from zero to sixty. The lesbians and gay men tend to have the fewest encounters per month, averaging seven and six, respectively. The women also report masturbating an average of five times per month, the men twelve. Of all subgroups, gay men masturbate the most, averaging sixteen times per month.

When it comes to the total number of sexual partners during their lifetimes, the variation is immense, ranging from one to five hundred. In general, the straight women have had the fewest, averaging thirteen, compared with sixteen for the lesbians and eighteen for the straight men. The bisexual women have had an average of twenty-four partners, sixty for the bisexual men. The gay men averaged 142 partners during their lifetimes, far more than any other subgroup. But judging from the lower numbers of recent encounters, it's reasonable to conclude that much of the gay men's sexual experimentation took place in the period before the AIDS epidemic struck.

Almost two-thirds of The Group are currently involved in primary relationships. The bisexuals—both men and women—are the most likely to be in such relationships, while the gay men are the least likely.

The vast majority of all members of The Group—87 percent, to be exact—say their self-esteem is moderate to very high.

5. Dr. Csikszentmihalyi (1990), pp. 120–21, reminds us that "memory is the oldest mental skill from which all others derive, for, if we weren't able to remember, we couldn't follow the rules that make other mental operations possible. . . . All forms of mental flow [his term for peak experience] depend on memory, either directly or indirectly."

6. Maslow (1968), p. 81.

2 THE EROTIC EQUATION

1. Tripp (1975), p. 59.

2. Freud (1930), *Civilization and Its Discontents.*

3. Studies have shown that women are often not as aware as men of when they are aroused. Summarizing this research, Singer (1984), p.

230, concludes, "Men's subjective reports of arousal agree well with physiological measures [of erection] . . . Women, on the other hand, often fail to report arousal, even at maximum physiological response, and under some conditions may report high arousal at low physiological levels."

4. Women who are interested in learning about using masturbation as a tool for self-awareness should see Betty Dodson's *Sex for One* (1987). Dodson is the best-known and most outspoken advocate of what she calls "selfloving."

5. Reik (1941), pp. 190, 96.

6. Tennov (1979), p. 23.

7. In *About Love* Solomon (1988) emphasizes the element of choice in love and downplays the experience of being overwhelmed by external forces. To me, one of the great and insoluble paradoxes of love is the fact that it is perhaps the most passive *and* active of all human experiences.

8. Person (1988), p. 138.

9. Ibid., p. 88.

10. Dr. Jim Weinrich, a sexologist at the University of California, San Diego, has focused on this variability of limerent attractions (compared to lusty ones) to help explain why bisexuality is apparently more common among women than among men. Because women are more inclined toward limerent attractions, and because limerent attractions tend to be more flexible when it comes to the gender of their objects, he reasons that women are more likely to respond, at one time or another, to both sexes. Weinrich (1987), see Chapter 6.

11. The concept of crystallization actually comes from Stendhal, the French novelist who, in 1822, wrote a highly personal collection of essays on passionate love. Ever since he has been rightly viewed as one of the seminal thinkers on the subject.

12. Tennov (1979), p. 26.

13. Ibid., p. 67.

3 FOUR CORNERSTONES OF EROTICISM

1. I am not suggesting that the cornerstones are the *only* sources of arousal-enhancing obstacles. I have no doubt that there are other ways of categorizing the universal life experiences that find expression in erotic life. I have been impressed by the lively discussions of other possi-

ble cornerstones whenever I speak to groups on the subject. It is precisely this type of dialogue that I wish to stimulate.

4 EMOTIONAL APHRODISIACS

1. It's often a mistake to interpret a cool, detached stance in sex as unemotional. An indifferent exterior may conceal a sense of pride about being able to remain in control (often seen as a sign of masculinity). At other times detached indifference is an effective means to communicate hostility toward the partner, or to cover up vulnerability. The most important feelings aren't necessarily the ones that show.

2. Quoted in Hillman (1960), p. 40.

3. Ibid., p. 43.

4. Hillman (1960), p. 97.

5. Reiss (1986).

6. Green (1977).

7. The fact that fear can produce involuntary arousal often contributes to the devastation of rape. Sometimes the terrified victim, whether female or male, is also sexually aroused, creating confusion or guilt. He or she may wonder in moments of shame and doubt, "Deep down, did I really *want* this to happen?" Many rapists and defense attorneys are aware of this fact and use it to transfer blame to the victims of sexual assault. Those who believe that a man can't be forced to have sex (especially intercourse) against his will are mistaken. In some instances the intense fear of the assault produces an unwanted erection.

8. Tennov (1979).

9. Ibid., p. 48.

10. Ibid., p. 54.

11. Apter (1992), p. 26.

12. For an insightful analysis of how hidden guilt can compel us to deprive ourselves of happiness, see Engel and Ferguson's *Imaginary Crimes* (1990).

13. Stoller (1979), p. 123.

14. Tavris (1989), p. 47.

15. Friday (1980), pp. 21–22.

5 YOUR CORE EROTIC THEME

1. Simon and Gagnon (1984).

2. I owe an enormous debt to two master psychologists whose similar lines of thought have contributed a great deal to my work: Robert Stoller and John Money. Stoller (1979), p. xi, speaks of "a *paradigmatic erotic scenario*—played out in daydream, or in choice of pornography, or in object choice, or simply in actions (such as styles of intercourse)—the understanding of which will enable us to understand the person." This "paradigmatic erotic scenario" is, in most respects, similar to what I call the CET. But it is a rather awkward phrase.

In a similar vein, Money (1986), p. 290, proposes the term "lovemap," which he defines as "a developmental representation or template in the mind and in the brain depicting the idealized lover and the idealized program of sexuoerotic activity projected in imagery or actually engaged in with that lover." Obviously, lovemaps are also like CETs. Unfortunately, since many CETs have little to do with love, the connotations of "lovemap" make the term too narrow for our purposes.

3. Dr. Stoller (1979), pp. 14–15, says that "secrecy is built into the sexual scripts . . . not only are secrets preserved, but the manner in which the secret is preserved is also a secret."

4. Simon and Gagnon (1984), p. 58.

5. Freud (1900), p. 608.

6. Thematic analysis has not progressed to the point where CETs can be categorized and counted. In fact, I doubt that phenomena as varied as erotic scenarios can ever be meaningfully categorized. Therefore, I have no way of proving which types of CETs are the most common, even among The Group. In many cases the operative CET (if any) was neither obvious nor discussed by the respondents.

7. Heiman (1975), p. 92.

8. Men who primarily rely on porn for masturbatory stimulation sometimes stop paying attention to their own fantasies. Their turn-ons are focused solely on sexual acts and body parts. Some of these men may have trouble recognizing the internal content of their turn-ons. Video porn, now the most popular medium for sexual imagery, often has the paradoxical effect of lulling the erotic imagination at the same time that it is enormously efficient at generating arousal and orgasm.

9. Those in search of erotic stories might start with one of Nancy Friday's collections of fantasy. *My Secret Garden* (1973) contains women's fantasies and *Men in Love* (1980) focuses on men's. A unique

and beautiful celebration of erotic photos, stories, and poetry can be found in David Steinberg's *Erotic by Nature* (1988). The "Sexuality Library" is an extensive catalogue of books about sex and a selection of videos concisely described for people of all sexual persuasions. Order the catalogue by writing to the Sexuality Library, 938 Howard Street, Suite 101, San Francisco, CA 94103.

10. Quoted in Zweig and Abrams (1991), p. 4

11. Stoller (1979), p. 6.

12. Moore (1990), p. 2.

13. Ibid., p. 85.

14. Stoller (1979), p. 125.

15. See Smirnoff (1970) for an analysis of "the masochistic contract."

16. Simon and Gagnon (1984), p. 56.

6 WHEN TURN-ONS TURN AGAINST YOU

1. Masters and Johnson proposed a four-phase model of sexual response consisting of excitement, plateau, orgasm, and resolution. Most sex therapists now follow the "triphasic" model proposed by Kaplan (1979), consisting of desire, excitement, and orgasm. This three-phase model recognizes the importance of desire (which Masters and Johnson *assumed* would be there if everything else was working) and helps clients and their therapists alike to define sexual problems more precisely.

2. All the stories are based on real people. But since therapy is confidential, I have carefully altered all details that could possibly reveal any person's identity. In addition, because different kinds of people grapple with similar problems, many of the stories are influenced by the experiences of more than one person. Not only does this approach help to conceal individual identities even further, it also allows each story to convey as much information as possible.

3. Like most therapists I prefer to stick with either individual or couple's therapy. There are times, however, when couple's therapy is indicated but clients are unable to explore crucial aspects of their turns-ons in the presence of their partners. Some individual meetings are often necessary to get at this material. As part of couple's therapy I'll occasionally meet individually with each partner in addition to our joint sessions. In Brian's case we started out with individual work because he knew he couldn't be fully honest with his partner.

Many therapists would never consider switching from individual to couple's therapy midstream. Frankly, it is a risky transition that doesn't always work. First, there is the danger that the incoming partner will feel left out or fearful that the therapist will side with the partner he or she already knows. Then there's the risk that the partner who's done the individual work will be locked in the "identified patient" role, that is, the one with the problem. Luckily, neither of these concerns became an issue in this case—and they usually don't if they are addressed openly.

4. The recovery movement is quite justified in its conviction that once a person is addicted, trying to figure out why can easily become a diversion. At least at first, all that matters is that the addict stops using. As sobriety takes hold, however, I believe that recovery requires courageous self-exploration. For many people the twelve steps of AA provide a structure for this process. But Nancy made an important observation: the vast majority of recovering addicts do not discuss the erotic functions that their chemical of choice once served, nor do they address the sexual implications of sobriety. Experts in addiction recovery pay a great deal of attention to preventing relapses, but the erotic dimensions of relapse are almost totally ignored.

5. Throughout our therapy there was never any indication that she had unresolved Oedipal attachments to her father, that is, wanting to possess him sexually as a child. Her pain had to do with her father's emotional unavailability.

6. Money (1986).

7. Monkeys that are prevented from sexual rehearsal play (e.g., simulated mounting of each other) are completely unable to function sexually as adults. Money (1980).

7 SEX AND SELF-HATE

1. People dissociate from trauma in many ways, including using sounds, images, or repeated thoughts to distract themselves. Some people describe "leaving their bodies" and observing their abuse with detached indifference, almost as if it wasn't really happening to them. The most serious form of dissociation is multiple personality disorder (MPD), in which the self splits into two or more "alter-egos." It's as if the mind is saying, "The horror of what happened is too much for one person to handle." MPD is the rarest form of dissociation.

2. Because people often dissociate while being abused, their memories may be stored differently than under normal circumstances. Memory

works through association; we are most likely to recall events when a current experience stimulates a related recollection. Abuse memories are often triggered by whatever the person was concentrating on at the time of the abuse, for example, an image, sound, or smell. Consequently, the triggering associations may at first seem strangely unrelated to the abuse.

3. For a balanced and sensitive exploration of questions related to repression, memory, and false memory, see Loftus (1993).

4. How unpleasant stimuli are converted into pleasurable ones is explored in "opponent-process theory" proposed by Solomon (1980).

5. Dr. John Money, the best-known student of paraphilias, has identified dozens of distinct erotic rituals that he places in six categories. See his book *Lovemaps* (1986).

6. I've worked with many clients, primarily men, driven by severe sexual compulsions who found significant relief with the help of one of the newer group of antidepressants (Prozac is the best known) that increase concentrations in the brain of the neurotransmitter serotonin. Sexually compulsive clients, particularly those who are also depressed, should be told about the possible benefits of combining medication with psychotherapy.

7. *Webster's New World Dictionary* (1988).

8. As a result of my experience with Nick and other clients, where there turned out to be an erotic dimension to seemingly nonsexual problems, I now invite clients to give some thought to their eroticism regardless of their reasons for entering therapy. The effects of eroticism extend throughout a person's life. I think it's a mistake to leave it to clients to find this out for themselves.

8 WINDS OF CHANGE

1. Defining the self is among the most difficult problems in psychology. On the one hand, we often speak of different selves—"true self," "false self," "shadow self," "old self," "new self," and so on. Carl Jung conceptualized the self as the totality of the person, encompassing conscious and unconscious aspects, as well as past, present, and future. This total self is also an inner blueprint for "individuation," his term for self-actualization. When referring to this total Self, I capitalize it.

2. For an insightful look at the healing power of compassion, see *Compassion and Self-Hate* by Theodore Rubin (1975).

3. When most men feel themselves about to ejaculate sooner than they want, their natural tendency is to tighten their muscles and hold their breath, as if trying to restrain themselves through sheer force. Ironically, this approach actually triggers the ejaculation reflex even faster. Deep, slow breathing, combined with muscle relaxation, is the most effective way to back away from the brink.

4. The best book about male sexuality, complete with specific exercises, is Zilbergeld's *The New Male Sexuality* (1992). Women wishing to use self-touch to help them learn how to have orgasms should read Barbach's *For Yourself: The Fulfillment of Female Sexuality* (1975). Barbach's *For Each Other* (1982) applies a similar approach to improving the quality of couple's sexual interaction. Those who want to learn more about erotic massage will enjoy Stubb's *Romantic Interludes: A Sensuous Lovers Guide* (1988).

9 LONG-TERM EROTIC COUPLES

1. Tripp (1975), p. 55.

2. In *The Couple's Journey* Susan Campbell (1980) spells out five stages: romance, power struggle, stability, commitment, and cocreation. Although it is oriented toward opposite-sex couples, same-sex couples may also find this system valuable. David McWhirter and Andrew Mattison (1984) offer a more complex analysis of long-term gay male relationships in *The Male Couple*.

3. "Deficiency-love" and "being-love" are part of a larger distinction Maslow makes between "deficiency motivation" and "growth motivation." See Maslow (1968), Chapter 3.

4. See *American Couples* by Blumstein and Schwartz (1983), p. 196.

5. Overall, gay men maintained the highest levels of sexual activity regardless of how long their relationships lasted. But over time they tended to have a greater proportion of sex outside their primary relationships, typically with casual partners.

6. Blumstein and Schwartz (1983), pp. 279–85.

7. Ibid., p. 295.

8. Many men hound themselves about lasting longer because their partners don't orgasm during intercourse even when, like Sandra, the partners express a clear need for direct clitoral stimulation. The pressure that so many men feel to "make" their lovers come with intercourse is a key reason that rapid ejaculation is the most common sexual concern

among straight men. Gay men, on the other hand, have this problem relatively rarely because no similar pressures exist between them.

9. "Lesbian bed death" refers to the tendency among many lesbians to become so focused on intimacy and unity that all passion-inspiring contrasts are extinguished, and with them their sexual desire. Another contributing factor is the training women receive not to initiate sex, waiting instead for a man to do it—obviously an unworkable proposition in a lesbian relationship. Erotic lesbian couples must resolve these two issues.

10. In her book *The Lesbian Erotic Dance* JoAnn Loulan (1990), probably the best-known lesbian sex therapist in the world, makes some crucial points about "butch" or "femme" roles among lesbians. First, she disputes the lesbian orthodoxy popular in the 1970s and 1980s that playing with gender roles is merely mimicking heterosexuality. She insists that opposite sexual "energies"—call them yin and yang if you like—are not strictly tied to male and female, but can be readily seen in human interactions of all kinds. She goes on to challenge the belief that these roles should be eradicated from lesbian sex and that everyone should strive for androgyny—an everyone-alike blending of butch and femme. Finally, she insists that when both partners strive to "act the same" they actually hasten the diminution of desire. Loulan encourages her readers to play with sex roles for erotic contrast and fun.

10 SIGNPOSTS TO EROTIC HEALTH

1. Student leaders at a well-known college have recently tried to banish uncertainties about consent with a policy requiring students to receive explicit verbal permission at each level of sexual involvement (e.g., "May I touch your breast?" or "May I stroke your thigh?"). Despite well-meaning intentions, the unspoken message is that the participants are too thoughtless or stupid to cope with the ambiguities of consent on their own. Heaven help us if such policies don't crumble under the weight of their own absurdity because their ultimate effect will be to reduce responsible decision making rather than to foster it.

2. King James Version, Matthew 5:28.

3. I'm not a biblical scholar, but perhaps the unattainability of moral perfection is Jesus' underlying point.

4. Stoller (1979), p. 136.

5. See Lynn Leight's *Raising Sexually Healthy Children* (1990), Sol and Judith Gordon's *Raising a Child Conservatively in a Sexually Permissive World* (1989), or Mary Calderone and Eric Johnson's *The*

Family Book About Sexuality (1989). If you're already familiar with Bernie Zilbergeld's *The New Male Sexuality* (1992) be sure to read the last chapter, "What You Can Do for Your Son." The best part of Nancy Friday's latest collection of fantasies, *Women on Top* (1991), is Part II: "Separating Sex and Love: In Praise of Masturbation." It is a passionate plea for mothers to allow their daughters to discover the power that comes through intimate knowledge of their own bodies.

11 EROS FULFILLED

1. Professionals interested in sexual trance phenomena can find a succinct summary and a good set of references in Swartz (1994). Donald Mosher's writings on "involvement theory" are particularly relevant (e.g., Mosher, 1980).

2. Reich was one of Freud's colleagues before they had a falling out. He emphasized the role of the body in repression, whereas Freud saw it as mostly a psychological phenomenon. In *The Function of the Orgasm*, Reich (1942), p. 270, said that "sexual life energy can be bound by chronic muscular tensions. Anger and anxiety can also be blocked by muscular tensions."

3. Maslow (1968), p. 94.

4. Unfortunately, when drug use becomes problematic, denial is normally part of the picture. We can easily convince ourselves that using alcohol or another drug before sex is "no big deal" even when we have become completely dependent. Therefore, it is wise to listen to observations of those who know and care about you. If someone close to you expresses a concern about your alcohol or drug use, take the concern very seriously, perhaps by seeking a professional evaluation. Left untreated, substance abuse wreaks havoc with a person's sex life.

5. In her book *The Space Between Us* Ruthellen Josselson (1992), p. 167, makes a crucial point about the importance of mutuality for psychological well-being: "At the opposite pole from mutuality and resonance are dissonance and emotional isolation, experiences that the psychological literature discusses more and understands better. In fact, psychopathology can be seen as the story of failed mutuality."

6. Carl Jung believed that experience of wholeness, so central to transcendence, includes a conscious awareness of the "archetypes" of the "anima" (the feminine aspect of a male) and "animus" (the masculine aspect of a female). According to Jung, this integration is a necessary challenge of personal growth.

7. Csikszentmihalyi (1990), p. 41.

8. Those who wish to apply structured rituals to the enhancement of sexual experience can find an accessible collection of exercises, along with a comprehensible rationale, in Margo Anand's *The Art of Sexual Ecstasy* (1989). Similar guidance can be found in the writings of Chinese Taoism. David Steinberg's *The Erotic Impulse* (1992) contains a broad sampling of writings on erotic transcendence. For those who live on either coast of the United States or in some urban centers of Europe, a number of workshops and seminars are also available on these subjects.

RECOMMENDED READINGS

Following are brief descriptions of seven books that are particularly relevant to the psychology of eroticism and peak experiences. Additional suggested readings can be found on specific topics in the Notes section. Look for entries preceded by the symbol.

The Homosexual Matrix (2nd ed.). C. A. Tripp. New York: New American Library, 1975, 1987.

This is the book that introduced me to the idea that overcoming obstacles intensifies arousal—the foundation of the paradoxical view of eroticism. It's far and away the best book available about homosexuality, though its insights cover much broader territory. See, for example, Chapter 4, "The Origins of Heterosexuality." But savor the entire masterpiece.

Sexual Excitement: Dynamics of Erotic Life. Robert Stoller. New York: Simon and Schuster, 1979.

Expanding psychoanalytic thinking deep into the erotic realm, Stoller proposes, among other things, that undercurrents of hostility fuel sexual excitement for most of us. Unfortunately, the book is illustrated by just one in-depth case. But it's loaded with understandable concepts about the unconscious sources of eroticism.

Love and Limerence: The Experience of Being in Love. Dorothy Tennov. Chelsea, Mich.: Scarborough House, 1979.

Combining a host of fascinating written materials and interviews, Tennov offers one of the most comprehensive looks at the overwhelming preoccupation of "limerence," her term for the intensity of early romance. Typically a mixture of intense pleasure and considerable distress, limerent experiences offer us another perspective on the paradoxical nature of eros.

Lovemaps. John Money. New York: Irvington, 1986.

Money's concept of the lovemap overlaps with what I've called the core erotic theme. He too is interested in the patterns that shape eroticism, but emphasizes "paraphilic lovemaps"—superfocused, problematic sexual rituals. It's too bad Money writes in such a difficult style, because this is a seminal work. Primarily for professional readers.

The Dangerous Edge: The Psychology of Excitement. Michael J. Apter. New York: Free Press, 1992.

Although Apter touches on sexuality only in passing, his ideas about our need for excitement, and the delicate relationship between excitement and anxiety, are directly applicable to eroticism. Among other topics, he outlines the conditions under which we are likely to enjoy being on the dangerous edge and why some people get into trouble with their excitement-seeking behaviors.

Toward a Psychology of Being. Abraham H. Maslow. New York: D. Van Nostrand, 1968.

Flow: The Psychology of Optimal Experience. Mihaly Csikszentmihalyi. New York: HarperPerennial, 1990.

These two books also mention sex only occasionally, but they offer the most insightful explorations of peak experiences available anywhere. Maslow's is the classic, which not only helped launch the field of humanistic psychology but also stimulated many researchers, myself included, to pay as much attention to our best experiences as we do to our troublesome ones.

Csikszentmihalyi's is a readable discussion of his clever research on the keys to happiness and vitality. Not surprisingly, many of the characteristics of "flow" are clearly evident in peak erotic experiences.

BIBLIOGRAPHY

Anand, Margo (1989). *The art of sexual ecstasy: The path of sacred sexuality for western lovers*. Los Angeles: Tarcher.

Apter, Michael J. (1992). *The dangerous edge: The psychology of excitement*. New York: Free Press.

Barbach, Lonnie (1975). *For yourself: The fulfillment of female sexuality*. New York: Doubleday.

——— (1982). *For each other: Sharing sexual intimacy*. New York: Signet.

de Beauvoir, Simone (1952). *The second sex*. New York: Bantam Books.

Blumstein, Philip, and Pepper Schwartz (1983). *American couples: Money, work, sex*. New York: Pocket Books.

Calderone, Mary, and Eric Johnson (1989). *The family book about sexuality*. New York: Harper and Row.

Campbell, Susan (1980). *The couple's journey: Intimacy as a path to wholeness*. San Luis Obispo, Calif.: Impact Publishers.

Csikszentmihalyi, Mihaly (1990). *Flow: The psychology of optimal experience*. New York: HarperPerennial.

Dodson, Betty (1987). *Sex for one: The joy of selfloving*. New York: Harmony Books.

Engel, Lewis, and Tom Ferguson (1990). *Imaginary crimes: Why we punish ourselves and how to stop*. Boston: Houghton Mifflin. (Also published as *Hidden Guilt*. New York: Pocket Books.)

Freud, Sigmund (1900). *The interpretation of dreams* (standard edition, trans. by James Strachey). New York: Norton.

——— (1910). "A Special Type of Object Choice Made by Men," in *Sexuality and the psychology of love*. New York: Collier.

——— (1912). "The Most Prevalent Form of Degradation in Erotic Life," in *Sexuality and the psychology of love*. New York: Collier.

——— (1930). *Civilization and its discontents* (standard ed., trans. by James Strachey). New York: Norton.

Friday, Nancy (1973). *My secret garden: Women's sexual fantasies*. New York: Pocket Books.

——— (1980). *Men in love: Men's sexual fantasies: The triumph of love over rage*. New York: Dell.

——— (1991). *Women on top: How real life has changed women's sexual fantasies*. New York: Pocket Books.

Gordon, Sol, and Judith (1989). *Raising a child conservatively in a sexually permissive world*. New York: Simon and Schuster.

Green, Elmer and Alyce (1977). *Beyond biofeedback.* New York: Delta Books.

Heiman, Julia R. (1975). "The physiology of erotica: women's sexual arousal." *Psychology Today 8* (April): 90–94.

Hillman, James (1960). *Emotion: A comprehensive phenomenology of theories and their meanings for therapy.* Evanston, Ill.: Northwestern University Press.

Josselson, Ruthellen (1992). *The space between us: Exploring the dimensions of human relationships.* San Francisco: Jossey-Bass.

Jung, Carl (1971). *The portable Jung.* New York: Viking Press.

Kaplan, Helen S. (1979). *Disorders of sexual desire.* New York: Brunner/Mazel.

Leight, Lynn (1990). *Raising sexually healthy children.* New York: Avon Books.

Loftus, Elizabeth F. (1993). "The reality of repressed memories." *American Psychologist 5:* 518–37.

Loulan, JoAnn (1990). *The lesbian erotic dance: Butch, femme, androgyny and other rhythms.* San Francisco: Spinsters Book Company.

McWhirter, David P., and Andrew Mattison (1984). *The male couple: How relationships develop.* Englewood Cliffs, N.J.: Prentice-Hall.

Maslow, Abraham H. (1968). *Toward a psychology of being.* New York: D. Van Nostrand.

——— (1971). *The farther reaches of human nature.* New York: Viking Press.

Masters, William H., and Virginia Johnson (1970). *Human sexual inadequacy.* Boston: Little, Brown.

Money, John (1980). *Love and love sickness: The science of gender difference and pair-bonding.* Baltimore: Johns Hopkins University Press.

——— (1986). *Lovemaps.* New York: Irvington.

Money, John, and Margaret Lamacz (1989). *Vandalized lovemaps: Paraphilic outcome of seven cases in pediatric sexology.* Buffalo, N.Y.: Prometheus.

Moore, Thomas (1990). *Dark eros: The imagination of sadism.* Dallas: Spring Publications.

Mosher, Donald L. (1980). "Three dimensions of depth of involvement in human sexual response." *Journal of Sex Research* 30 (1): 1–42.

Paglia, Camille (1990). *Sexual personae: Art and decadence from Nefertiti to Emily Dickinson.* New York: Vintage Books.

Person, Ethel S. (1988). *Dreams of love and fateful encounters: The power of romantic passion.* New York: Norton.

Reich, Wilhelm (1942, 1973). *The function of the orgasm* (trans. by Vincent Carfagno). New York: Farrar, Straus and Giroux.

Reik, Theodor (1941). *On love and lust: On the psychoanalysis of romantic and sexual emotions.* New York: Grove Press.

Reiss, Ira L. (1986). *Journey into sexuality: An exploratory voyage.* Engelwood Cliffs, N.J.: Prentice-Hall.

Rubin, Theodore I. (1975). *Compassion and self-hate: An alternative to despair.* New York: Collier Books.

Schnarch, David M. (1991). *Constructing the sexual crucible: An integration of sexual and marital therapy.* New York: Norton.

Simon, William, and John H. Gagnon (1984). "Sexual scripts." *Society* (November/December): 53–60.

Singer, Barry (1984). "Conceptualizing sexual arousal and attraction." *Journal of Sex Research* 20: 230–40.

Smirnoff, V. N. (1970). "The masochistic contract." *International Journal of Psychoanalysis* 50: 666–70.

Solomon, R. L. (1980). "The opponent-process theory of acquired motivation." *American Psychologist* 35: 691–712.

Solomon, Robert C. (1988) *About love.* New York: Touchstone.

Steinberg, David, ed. (1988). *Erotic by nature: A celebration of life, love, and of our wonderful bodies.* North San Juan, Calif.: Shakti Press.

——— (1992). *The erotic impulse: Honoring the sensual self.* Los Angeles: Tarcher.

Stendhal (1822). *Love* (trans. by Gilbert and Suzanne Sale, 1975). Middlesex, England: Penguin Books.

Stoller, Robert J. (1964). "A contribution to the study of gender identity." *International Journal of Psychoanalysis* 45: 220–26.

——— (1979). *Sexual excitement: Dynamics of erotic life.* New York: Simon and Schuster.

Stubbs, Kenneth R. (1988). *Romantic interludes: A sensuous lovers guide.* Larkspur, Calif.: Secret Garden.

Swartz, Louis H. (1994). "Absorbed states play different roles in female and male sexual response: Hypotheses for testing." *Journal of Sex and Marital Therapy* 20 (3): 244–53.

Tavris, Carol (1989, rev. ed.) *Anger: The misunderstood emotion.* New York: Touchstone.

Tennov, Dorothy (1979). *Love and limerence: The experience of being in love.* Chelsea, Mich.: Scarborough House.

Tripp, C. A. (1975, 1987, 2nd ed.) *The homosexual matrix.* New York: New American Library.

Weinrich, James D. (1987). *Sexual landscapes: Why we are what we are, why we love whom we love.* New York: Charles Scribner's Sons.

Zilbergeld, Bernie (1992). *The new male sexuality.* New York: Bantam Books.

Zweig, Connie, and Jeremiah Abrams, eds. (1991). *Meeting the shadow: The hidden power of the dark side of human nature.* Los Angeles: Tarcher.

INDEX